Springer Theses

Recognizing Outstanding Ph.D. Research

For further volumes:
http://www.springer.com/series/8790

Aims and Scope

The series "Springer Theses" brings together a selection of the very best Ph.D. theses from around the world and across the physical sciences. Nominated and endorsed by two recognized specialists, each published volume has been selected for its scientific excellence and the high impact of its contents for the pertinent field of research. For greater accessibility to non-specialists, the published versions include an extended introduction, as well as a foreword by the student's supervisor explaining the special relevance of the work for the field. As a whole, the series will provide a valuable resource both for newcomers to the research fields described, and for other scientists seeking detailed background information on special questions. Finally, it provides an accredited documentation of the valuable contributions made by today's younger generation of scientists.

Theses are accepted into the series by invited nomination only and must fulfill all of the following criteria

- They must be written in good English.
- The topic should fall within the confines of Chemistry, Physics, Earth Sciences and related interdisciplinary fields such as Materials, Nanoscience, Chemical Engineering, Complex Systems and Biophysics.
- The work reported in the thesis must represent a significant scientific advance.
- If the thesis includes previously published material, permission to reproduce this must be gained from the respective copyright holder.
- They must have been examined and passed during the 12 months prior to nomination.
- Each thesis should include a foreword by the supervisor outlining the significance of its content.
- The theses should have a clearly defined structure including an introduction accessible to scientists not expert in that particular field.

Elise Jennings

Simulations of Dark Energy Cosmologies

Doctoral Thesis accepted by
Durham University, UK

 Springer

Authors
Dr. Elise Jennings
Durham University
Durham
UK

Supervisors
Prof. Dr. Carlton M. Baugh
Durham University
Durham
UK

Dr. Silvia Pascoli
Durham University
Durham
UK

ISSN 2190-5053 ISSN 2190-5061 (electronic)
ISBN 978-3-642-43640-6 ISBN 978-3-642-29339-9 (eBook)
DOI 10.1007/978-3-642-29339-9
Springer Heidelberg New York Dordrecht London

For my brother David

Supervisors' Foreword

The discovery of the accelerating expansion of the Universe in the mid 1990s was largely unexpected. In a matter-dominated universe, the expansion slows with time. Observations of distant supernovae were found to be consistent with the distance-redshift relation expected in a cosmology with a dynamically dominant dark energy. In such a universe, the rate of expansion is speeding up. This breakthrough was marked by the award of the 2011 Nobel Prize in Physics to the two observational teams that carried out the supernova surveys.

The realisation that the expansion of the universe is speeding up rather than slowing down has galvanised the astronomical community into finding ways to quantify the nature of the dark energy. A series of ambitious galaxy surveys is planned which will culminate with the launch of the European Space Agency's Euclid mission, currently scheduled for 2019, which aims to produce the most extensive galaxy map ever made.

Cosmological structure grows in a battle between gravity and the expansion of the Universe. Dark energy leaves an imprint on this process by influencing the cosmic expansion rate. Also, the large-scale distribution of matter contains characteristic scales which can be used as "standard rulers" with which to measure the distance-redshift relation and hence to constrain the values of the basic cosmological parameters. The growth of cosmic structure can be modelled under special conditions using analytical methods. The final stages of the collapse of density perturbations are, however, highly nonlinear with significant "cross talk" between different scales. The cosmologists' tool of choice to model gravitational instability is computer simulation. The most popular technique is N-body simulation, in which the density of the universe is represented in a coarse-grained approximation by particles. These particles are many orders of magnitude more massive than the particle physics candidates for the dark matter, with a mass determined by the problem under consideration and the computing resources available. The signatures of dark energy models on the galaxy distribution are only subtly different, so it is essential that accurate theoretical predictions are developed to interpret forthcoming galaxy surveys, which means that N-body simulations of structure formation in competing cosmologies are needed.

This thesis sets out the state of the art in using computer simulation to model the growth of large-scale structure in the matter distribution in different dark energy cosmologies. For the first time, an accurate model for the equation of state of the dark energy is implemented over the full range of epochs modelled in the simulation, which can be identified with the properties of the underlying field. The impact of small but appreciable amounts of dark energy at early epochs is taken into account, as reflected in the initial fluctuation spectrum used in the calculations.

The key result of the thesis is that the large-scale clustering of matter is much more complicated than is typically assumed in forecasts for future observations. In particular, the impact of peculiar motions, departures from the Hubble expansion induced by the lumpiness of the matter distribution, is substantially different from that expected in traditional models. The application of such models to the simulation results can lead to fundamentally flawed conclusions about the behaviour of dark energy.

The thesis sets out a manifesto for the analysis of future galaxy surveys. These surveys will be so large that they will allow measurements of galaxy clustering on large scales of unprecedented accuracy. The author argues that it is essential that to use computer simulations to guide the development of improved analytical models which can capture the simulation results. This symbiosis between simulation and simple analysis will become an essential element of the quest to establish the nature of dark energy.

Durham, UK, February 2012

Prof. Carlton M. Baugh
Dr. Silvia Pascoli

Acknowledgments

First I would like to express my sincere thanks to Carlton and Silvia for the excellent supervision and guidance they have both given me. Throughout my Ph.D. our regular meetings were always enjoyable, inspiring and fruitful. Thank you both for your patience and invaluable suggestions. It has been a privilege for me to have worked on this topic with you for 3 years.

All of the work for this thesis was carried out on The Cosmology Machine, COSMA, in Durham. I would like to thank Lydia Heck, Alan Lotts, John Helly and Raul Angulo for all their help and guidance in running the simulations.

Thanks to all my friends and office mates in the department, especially to Violeta, Claudia, Nico, Alvaro, Nikos Fanidakis, Nikos Nikoloudakis, Gabriel, Dave, Alex and Eimear. Thank you all for making my Ph.D. so much fun. Thanks to Luke, Austin, Laya, Fruzana and Emma for so many nights out and parties that have undoubtly improved this thesis.

My thanks and love go to my brother David who has been the best brother, friend and parent to me. I can honestly say that without you I would probably be a very unhappy vet and that this thesis would never have been written.

I have kept my final and biggest thank you for Brian, who is the close friend I have ever had. Thank you so much for putting up with me during the stressful times of this thesis and for always encouraging, loving and supporting me.

Contents

Chapter 1
Introduction

The common assumption in modern cosmology is that the Universe is statistically homogeneous and isotropic, and therefore can be accurately described by the Friedmann-Robertson-Walker (FRW) metric which has one degree of freedom, the cosmic scale factor, $a(t)$. In the current cosmological model, our Universe has evolved from a homogeneous state after the big bang to a highly inhomogeneous state of galaxies and clusters of galaxies, the energy density of which is composed of 4 % baryons, 22 % dark matter and 74 % dark energy today (Sanchez et al. 2006; Komatsu et al. 2010). The success of the 'hot big bang' model is clear from observations such as the microwave background blackbody radiation from the early Universe and from predictions of light element abundances from big bang nucleosynthesis. Despite this, there remain several challenges which the model fails to overcome, such as the nature of the inflationary mechanism and the presence of dark matter and dark energy. The growth of large scale structure in the Universe is an extremely important tool which can be used to probe fundamental physics such as the nature of dark energy and the theory of gravity. Structure formation is driven by a competition between the expansion of the Universe and gravitational attraction. By measuring the rate at which overdensities grow and their clustering statistics we can test different cosmological models. This chapter reviews the growth of density perturbations, the evidence for the accelerating cosmic expansion and discusses viable models which can solve the dark energy problem. We also present an overview of current and future probes of dark energy and modified gravity. In the coming years, new galaxy surveys and other cosmological observations will provide very precise measurements of the properties of dark energy. The work presented in this thesis uses state of the art modelling of dark energy cosmologies to provide accurate theoretical predictions for several cosmological probes.

E. Jennings, *Simulations of Dark Energy Cosmologies*, Springer Theses,
DOI: 10.1007/978-3-642-29339-9_1, © Springer-Verlag Berlin Heidelberg 2012

1.1 The Growth of Linear Fluctuations

The FRW metric is given by

$$\mathrm{d}s^2 = -\mathrm{d}t^2 + a^2(t)[(1 - kr^2)\mathrm{d}r^2 + r^2\mathrm{d}\theta^2 + \sin^2\theta\,\mathrm{d}\phi^2], \qquad (1.1)$$

where $k = 0, \pm 1$ is a parameter which describes the spatial curvature, r, θ and ϕ are spherical coordinates and the parameter $a(t)$ acts as an evolutionary factor in the distance and is referred to as the scale factor or expansion factor. This can also be expressed as a function of redshift z, where $(1 + z) = 1/a(t)$. The evolution of $a(t)$ is described by the following two equations,

$$\frac{\dot{a}^2 + k}{a^2} = \frac{8\pi G\rho}{3}\,; \qquad\qquad \mathrm{d}(\rho a^3) = -P\mathrm{d}a^3, \qquad (1.2)$$

where ρ is the energy density of the Universe and P is the pressure. The second conservation equation can be combined with an equation of state, which relates the pressure and the energy density, to determine the evolution of energy density, $\rho(a)$. We can define a critical density for a flat Universe as

$$\rho_{\mathrm{crit}} = \frac{3H_0^2}{8\pi G} = 1.88h^2 \times 10^{-29}\mathrm{g/cm}^3, \qquad (1.3)$$

where $H = \dot{a}/a$ is the Hubble parameter, H_0 is its current value and $h = H_0/$ $(100\,\mathrm{km/s/Mpc})$. The dimensionless density parameter for a component, x, is defined as

$$\Omega_x = \frac{\rho_x(t_0)}{\rho_{\mathrm{crit}}(t_0)}, \qquad (1.4)$$

where $\rho_x(t_0)$ is the density today.

The comoving coordinate, $\vec{x}$, is given by the physical position, $\vec{r}$ multiplied by the cosmological scale factor, a, as $\vec{r} = a\vec{x}$. The physical velocity is then the sum of the Hubble expansion velocity and a peculiar velocity as $\vec{v} = \dot{\vec{r}} = \dot{a}\vec{x} + a\dot{\vec{x}}$. In this thesis we analyse the formation and evolution of dark matter overdensities defined as

$$\delta(x) \equiv \frac{\rho(x)}{\bar{\rho}} - 1, \qquad (1.5)$$

where bar denotes the unperturbed (i.e. homogeneous) matter density. In the Newtonian limit the evolution of first order cosmological matter perturbations is described by the linearized equations of motion in comoving units as

$$\text{Euler} \quad \dot{\vec{v}} + \vec{v} \cdot \nabla\vec{v} + 2\frac{\dot{a}}{a}\vec{v} = -\frac{\nabla P}{\rho} - \frac{\nabla\Phi}{a^2}, \qquad (1.6)$$

$$\text{continuity} \quad \dot{\delta} + \nabla \cdot [(1 + \delta)\vec{v}] = 0, \tag{1.7}$$

$$\text{Poisson} \quad \nabla^2 \Phi = 4\pi G \bar{\rho} a^2 \delta, \tag{1.8}$$

where Φ is the gravitational potential and G is Newton's gravitational constant. In the equations above, differentiation with respect to $\vec{x}$ is denoted by ∇ and with respect to time as a dot. We can combine the three equations above to obtain one equation which describes the growth of matter perturbations in an expanding universe. In a dust universe, i.e. one where $P = 0$, taking the divergence of the Euler equation, using the continuity equation to eliminate $\nabla \cdot \vec{v}$ and replacing $\nabla^2 \Phi$ using Poisson's equation gives the growth equation for density perturbations,

$$\ddot{\delta} + 2H\dot{\delta} = 4\pi G \bar{\rho} \delta. \tag{1.9}$$

The growth of large scale structure, described by Eq. 1.9, is determined by a competition between the attractive force of gravity, causing slightly denser regions to increase in density, and the expansion rate of the Universe. The expansion rate introduces an effective friction term into Eq. 1.9 corresponding to the Hubble drag term, H. The general solution to this equation can be written in terms of a growing and a decaying mode solution where, in a matter dominated Universe, the growing mode can be written as

$$\delta \propto D(t) \propto a, \tag{1.10}$$

where a is the scale factor. At early times, when the matter density perturbations are small and the density contrast $\delta(\vec{x}, t) \ll 1$, only the growing mode is present and the field grows self-similarly in time as

$$\delta(\vec{x}, t) = D(t)\delta_0(\vec{x}). \tag{1.11}$$

A statistical description of the inhomogeneities in a field is very useful as the distribution of matter in the Universe can vary from point to point with overdensities of different wavelengths and amplitudes. The matter overdensity in Fourier space is given as

$$\delta(\vec{k}) = (2\pi)^{-3/2} \int \mathrm{d}^3x \, \delta(\vec{x}) \, e^{i\vec{k}\cdot\vec{x}}. \tag{1.12}$$

In linear perturbation theory each k mode evolves independently resulting in the following correlation

$$\langle \delta(\vec{k})\delta(\vec{k}')\rangle = \langle |\delta(k)|^2 \rangle \delta^3(\vec{k} - \vec{k}') \equiv P(k)\delta^3(\vec{k} - \vec{k}'), \tag{1.13}$$

where $P(k)$ is the power spectrum. From Eq. 1.11 the power spectrum as a function of time in linear perturbation theory is separable as

$$P(k, t) = \frac{D(t)^2}{D(t_0)^2} P(k, t_0), \tag{1.14}$$

where $D(t_0)$ is the linear growth factor at the present epoch.

1.2 The Accelerating Expansion of the Universe

The discovery that the expansion of the Universe is accelerating was first made over ten years ago by two independent groups observing distant supernovae (Riess et al. 1998; Perlmutter et al. 1999). Type Ia supernova (SN) are white dwarf stars in a binary system which are accreting mass from a companion star. A thermonuclear reaction occurs when the white dwarf reaches its Chandrasekhar mass, $\sim 1.4 M_\odot$, resulting in a very bright outburst with typical peak luminosities a few billion times that of our Sun. Using an empirical relation between the peak luminosity and the rate at which the light curve decays, Type Ia SN are excellent 'standardizable' candles, providing a distance measure which can probe the expansion history of the Universe (e.g. Phillips 1993).

In 1998, the Supernova Cosmology Project (SCP) and the High-z SN Search Team (HZT) found that distant supernovae at $z \sim 0.5$ were about 0.2 magnitudes dimmer than expected. Early results could only constrain a linear combination of Ω_{m} and Ω_Λ, the dimensionless dark energy parameter today,o close to $\Omega_{\mathrm{m}} - \Omega_\Lambda$, even after quite restrictive assumptions e.g. priors on h and on the curvature Ω_{k}. These observations were the first concrete evidence of a non-zero positive Ω_Λ. The SCP (Perlmutter et al. 1999) analysed 42 Type Ia SN between redshifts 0.18 and 0.83 and were able to constrain the relation $0.8\Omega_{\mathrm{m}} - 0.6\Omega_\Lambda \approx -0.2 \pm 0.1$. These SN results had to be used in combination with other observations of the geometry of the Universe to give a detection of Ω_Λ. For a flat Universe ($\Omega_{\mathrm{m}} + \Omega_\Lambda = 1$) the SCP found $\Omega_{\mathrm{m}}^{\mathrm{flat}} = 0.28^{+0.09}_{-0.08}$.

Some doubts surrounded the robustness of these early SN measurements as it was suggested that host-galaxy extinction by a hypothetical grey dust could be obscuring the SN making them appear dimmer (Aguirre 1999). In the following decade, advances in instrumentation and improved host-galaxy extinction estimators have resulted in precise measurements which rule out dust extinction as an alternative to an accelerating expansion (Riess et al. 2004). Recently direct SN searches, for example with the Hubble space telescope (HST) (Knop et al. 2003), have obtained high quality light curves and are able to constrain the cosmological parameters independently of other datasets. Recent supernovae observations of the distance modulus versus redshift (see Appendix A.1) from the Supernova Legacy Survey (SNLS) (Astier et al. 2005) and the ESSENCE survey (Miknaitis et al. 2007; Wood-Vasey et al. 2007) are shown in Fig. 1.1.

Since these early SN measurements, the cosmic acceleration has been firmly established using robust independent evidence from the cosmic microwave background (CMB). The CMB is the relic radiation from the early Universe, emitted at a redshift

Fig. 1.1 The Hubble diagram for low redshift supernovae from Wood-Vasey et al. (2007). Residuals from an open cosmological model with $\Omega_m = 0.3$ and $\Omega_\Lambda = 0$ are shown in the *lower panel*. The *solid line* plotted is the best fit cosmology with $(w, \Omega_m, \Omega_\Lambda) = (-1, 0.27, 0.73)$. The *dotted* and *dashed lines* correspond to cosmologies with $(\Omega_m, \Omega_\Lambda)$ equal to $(0.3, 0.0)$ and $(1.0, 0.0)$ respectively

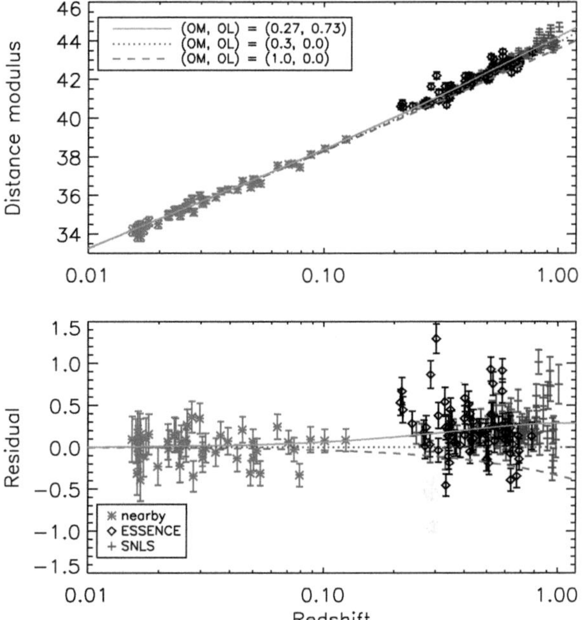

of $z \sim 1090$, when the ionized photon and electron plasma cooled, allowing neutral hydrogen to form. The photons then decoupled from the matter at what we refer to as the last scattering surface and have free streamed through the Universe with little subsequent interaction. As a result, the CMB is incredibly homogeneous with fluctuations in the temperature power spectrum of a few μK or, equivalently, at the level of 10^{-5} over the entire sky. Early measurements of the CMB with the COBE satellite reported the amplitude of the temperature fluctuations on large angular scales ($\theta > 7°$) and found the anisotropies to be consistent with Gaussian statistics and a scale invariant power spectrum (Bennett et al. 1996). The Wilkinson Microwave Anisotropy Probe (WMAP) (Komatsu et al. 2010) was launched in 2001 and has produced the first fine-resolution full-sky map of the CMB resulting in precision measurements of the temperature and polarisation power spectra (see Fig. 1.2).

The acoustic oscillations in the photon-baryon fluid before decoupling leave a characteristic imprint on the CMB with the first peak appearing on angular scales of about 1 degree. This corresponds to the sound horizon, r_s (see Appendix A.1) which is the maximum distance the sound wave could have travelled before decoupling. The apparent size of the sound horizon is sensitive to the spatial curvature of the Universe. By locating the first peak in the CMB power spectrum using the WMAP 7 year data, Komatsu et al. (2010) constrain the total density of the Universe to be $\Omega_{tot} = 1$ to better than 1 %. With several independent probes of the matter density finding $\Omega_m \sim 0.3$, this would imply a missing energy content of 70 %. In order to have the large scale structures we see today, such a component must have only emerged

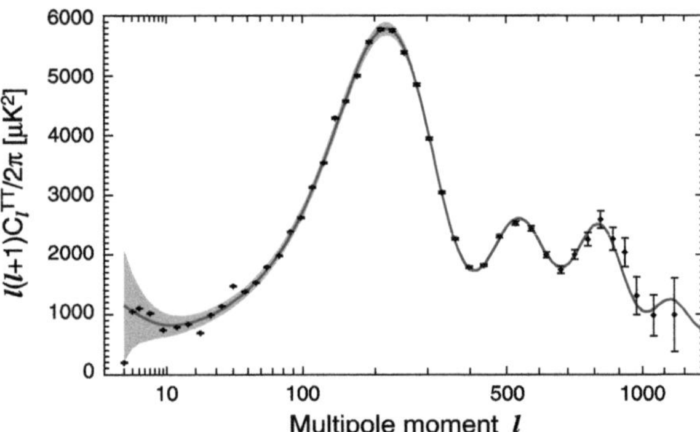

Fig. 1.2 The WMAP 7 year temperature power spectrum (points; Larson et al. 2010) showing the baryonic acoustic oscillations. The curve is the best fit to the data ΛCDM model with $\Omega_b h^2 = 0.02270$, $\Omega_c h^2 = 0.1107$ and $\Omega_\Lambda = 0.738$. The *grey* shaded region represents cosmic variance

recently to dominate the total energy density of the Universe, which constrains its equation of state parameter, the ratio of the pressure to the energy density of the fluid, $w = P/\rho \leq -1/3$. The spatially flat Universe implied by the CMB data also agrees with the predictions from theories of primordial inflation.

Further observational evidence for cosmic acceleration comes from measurements of large scale structure, for example combining the shape of the matter power spectrum with CMB data. Efstathiou et al. (2002) conducted a joint analysis of the power spectrum of the 2dF Galaxy Redshift Survey (2dFGRS) and the CMB spectrum and found $0.65 < \Omega_\Lambda < 0.85$ at 2σ uncertainty. The shape of the matter power spectrum is sensitive to the parameter combination $\Omega_m h$ while the CMB alone cannot constrain h or Ω_Λ but is sensitive to the combination of the physical densities $w_b = \Omega_b h^2$ and $w_c = \Omega_c h^2$ where Ω_b and Ω_c are the density parameters in baryons and cold dark matter respectively. Combining these two measurements helps to break parameter degeneracies and provides important constraints on cosmology which are independent of SN data, which, as discussed above, could be subject to possible systematic errors, e.g, the dependence on host galaxy properties and dust extinction.

A further probe of cosmology is the apparent size of the acoustic oscillations described above in the galaxy distribution. These features, called BAO, are weak in the matter distribution compared to their amplitude in the CMB power spectrum. This is because the total matter density exceeds the baryon density by a large factor, leading to BAO which are damped in amplitude (e.g. Meiksin et al. 2000). These delicate features can be erased by a number of dynamical and statistical effects as structure grows and galaxies form (Angulo et al. 2008). Nevertheless, the BAO have been detected in the low redshift galaxy distribution (Cole et al. 2005; Eisenstein et al. 2005). Figure 1.3 shows the two point correlation function of luminous red galaxies in the Sloan Digital Sky Survey (SDSS) (Eisenstein et al. 2005) with a bump occurring at $r \sim 110 h^{-1}$Mpc corresponding to the sound horizon. The CMB acts as a standard ruler allowing us to determine the spatial geometry at $z \sim 1090$, while the BAO provide a complementary ruler which to date has been measured at lower redshifts, $z \lesssim 1$. The apparent size of the BAO, given the measurement of the sound horizon scale from the CMB, allows us to constrain the distance to a given

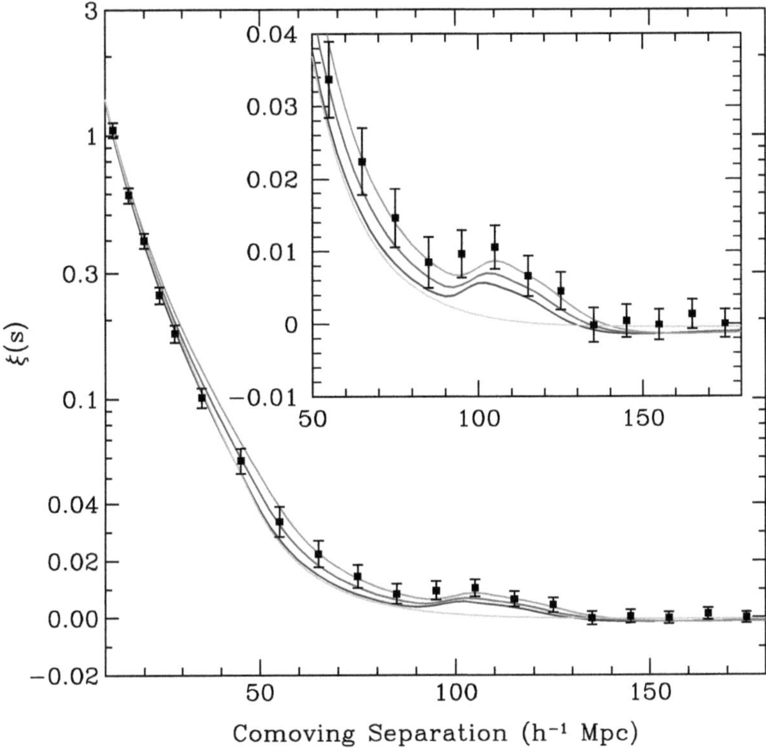

Fig. 1.3 The correlation function for SDSS luminous red galaxies with the BAO peak at $r \sim 110h^{-1}$Mpc (Eisenstein et al. 2005). The *lines* show different cosmologies with $\Omega_{\mathrm{m}}h^2 = 0.12$ (*top line*), 0.13 (*second line*) and 0.14 (*third line*) with $\Omega_{\mathrm{b}}h^2 = 0.024$ and $n_s = 0.98$ in all cases. The *bottom line* represents a pure cold dark matter model $\Omega_{\mathrm{m}}h^2 = 0.105$ with no acoustic peak

redshift and hence the cosmological world model (Hu and Haiman 2003; Blake and Glazebrook 2003).

The three probes discussed here, SNe, CMB and BAO, are complementary and constrain different regions of parameter space (see Fig. 1.4; Kowalski et al. 2008). Individual datasets are affected by different parameter degeneracies. For example, the WMAP data alone cannot constrain the spatial curvature but with two or more distance measurements it is possible to break the degeneracy between Ω_{k} and Ω_{m}. In fact, WMAP measurements together with BAO can completely fix Ω_{k} nearly independently of the dark energy equation of state (Komatsu et al. 2010). As CMB measurements are sensitive to the combination $\Omega_{\mathrm{m}}h^2$, a flatness prior together with constraints on h from HST (Knop et al. 2003) are used to break this degeneracy and obtain good constraints on $\Omega_{\Lambda} = 1 - (\Omega_{\mathrm{m}}h^2)/h^2$. It is the robustness of these results, together with additional probes such as the Integrated Sachs Wolfe effect, weak and strong gravitational lensing and X-ray clusters (which we discuss in more detail in

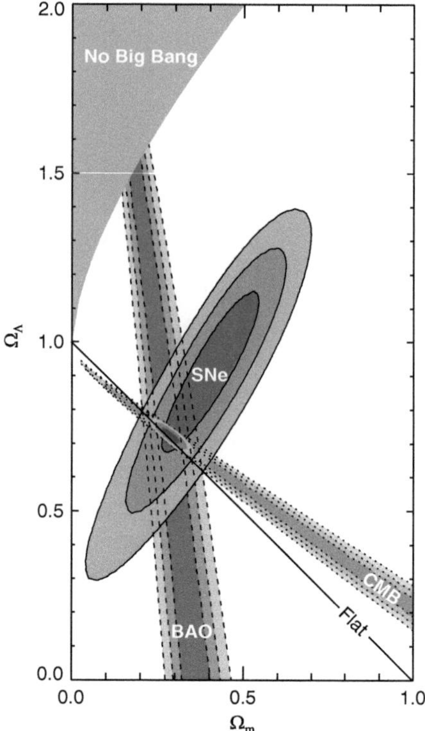

Fig. 1.4 The constraints on Ω_m and Ω_Λ from CMB, BAO and SN observations from Kowalski et al. (2008)

the following section) that have led to our current concordance cosmological model, where dark energy accounts for $\sim 70\%$ of the total energy density of the Universe.

1.3 Cosmological Models

The observed accelerating expansion of the Universe points towards new physics and explaining it is one of the biggest challenges in cosmology today. One explanation of the accelerating expansion of the Universe is that a negative pressure dark energy component dominates the present cosmic density (Sánchez et al. 2009; Komatsu et al. 2010). Examples of dark energy models include the cosmological constant and a dynamical scalar field such as quintessence (see e.g. Copeland et al. 2006 for a review). Other possible solutions require modifications to general relativity and include extensions to the Einstein–Hilbert action, such as $f(R)$ theories or braneworld cosmologies (see e.g. Dvali et al. 2000; Oyaizu 2008).

The concordance model, ΛCDM (cold dark matter and cosmological constant model), assumes a negative pressure component in the Universe acting as a fluid with a constant equation of state, $w = -1$, which drives the accelerated expansion.

The addition of a cosmological constant, Λ, to Einstein's theory of gravity is the most familiar and the simplest candidate for dark energy (see e.g. the review by Carroll 2001). Einstein's equation with a cosmological constant is given by

$$G_{\mu\nu} = R_{\mu\nu} - \frac{1}{2}g_{\mu\nu}R = 8\pi G T_{\mu\nu} + \Lambda g_{\mu\nu}, \tag{1.15}$$

where $R_{\mu\nu}$ and R are the Ricci tensor and scalar respectively, $T_{\mu\nu}$ is the energy-momentum tensor, G is Newton's constant and $g_{\mu\nu}$ is the space-time metric. Including a cosmological constant term modifies the RHS of Einstein's equation which is interpreted as adding a new fluid component to the Universe—referred to as 'dark energy'. This negative pressure component is generally assumed to be the vacuum energy arising from the zero point fluctuations of quantum fields. Despite the success of ΛCDM at fitting much of the available observational data (Sánchez et al. 2009), this model fails to address two important issues, the fine tuning problem and the coincidence problem. The fine-tuning problem arises from the vast discrepancy between the vacuum energy level predicted by particle physics, generically given by Λ^4, where Λ is the physics scale considered, and the value of missing energy density inferred cosmologically, $\rho \sim 10^{-47}$ GeV^4. In the standard model of particle physics, Λ could be at the Planck scale, giving $\Lambda \sim 10^{18}$ GeV. This leads to the famous 120 orders of magnitude difference between the measured energy density and the predicted zero point energy density of the Universe. The coincidence problem refers to the fact that we happen to live around the time at which dark energy has emerged as the dominant component of the Universe, and has a comparable energy density to matter, $\rho_{\mathrm{DE}} \sim \rho_{\mathrm{m}}$. It is a puzzle that we live in the brief intermediate phase when the matter density of the Universe is similar to the dark energy density given their different rates of evolution, $\rho_{\mathrm{m}} \propto a^{-3}$ and $\rho_\Lambda \sim$ constant (however see Bianchi and Rovelli 2010 for a recent discussion).

Quintessence models were devised to solve the fine tuning and coincidence problems of ΛCDM. In these models, the cosmological constant is replaced by an extremely light scalar field which evolves slowly (Ratra and Peebles 1988; Wetterich 1988; Caldwell et al. 1998; Ferreira and Joyce 1998). An abundance of quintessence models has been proposed in the literature which can resolve the coincidence problem and explain the observationally inferred amount of dark energy. Models of quintessence dark energy can have very different potentials, $V(\varphi)$, but can share common features. The potentials provide the correct magnitude of the energy density and are able to drive the accelerated expansion seen today. The form of the scalar field potential determines the trajectory of the equation of state, $w(z)$, as it evolves in time. Hence, different quintessence dark energy models have different dark energy densities as a function of time, $\Omega_{\mathrm{DE}}(z)$. This implies a different growth history for dark matter perturbations from that expected in ΛCDM.

Here we briefly review some general features of quintessence models; more detailed descriptions can be found, for example, in Ratra and Peebles (1988); Wetterich (1988); Ferreira and Joyce (1998); Copeland et al. (2006) and Linder (2008). The main components of quintessence models are radiation, pressureless matter and a quintessence scalar field, denoted by φ. This dynamical scalar field is a slowly evolving component with negative pressure. This multifluid system can be described by the following action

$$S = \int d^4x \sqrt{-g} \left(-\frac{R}{2\kappa} + \mathscr{L}_{m+r} + \frac{1}{2}\, g^{\mu\nu}\, \partial_\mu \varphi\, \partial_\nu \varphi - V(\varphi)\right), \qquad (1.16)$$

where R is the Ricci scalar, $\mathscr{L}_{m+r}$ is the Lagrangian density of matter and radiation, $\kappa = 8\pi G$, g is the determinant of a spatially flat FRW metric tensor $g_{\mu\nu}$ and $V(\varphi)$ is the scalar field potential. We assume that any couplings to other fields are negligible so that the scalar field interacts with other matter only through gravity. Minimising the action with respect to the scalar field leads to its equation of motion

$$\ddot{\varphi} + 3\,H\,\dot{\varphi} + \frac{dV(\varphi)}{d\varphi} = 0, \qquad (1.17)$$

where H is the Hubble parameter and we have assumed the field is spatially homogeneous, $\varphi(\vec{x}, t) = \varphi(t)$. The impact of the background on the dynamics of φ is contained in the $3H\dot{\varphi}$ term. Once a standard kinetic term is assumed in the quintessence model, it is the choice of potential which determines the equation of state w as

$$w = \frac{\dot{\varphi}^2/2 - V(\varphi)}{\dot{\varphi}^2/2 + V(\varphi)}. \qquad (1.18)$$

In general in these theories if the contributions from the kinetic ($\dot{\varphi} = 0$) and gradient energy ($d\varphi/d\vec{x} = 0$) are negligible, then the effect of the scalar field is equivalent to a cosmological constant which behaves as a perfect fluid, with $P = -\rho$ or $w = -1$. Specific classes of quintessence models are discussed in more detail in Chap. 3.

Einstein's theory of general relativity describes the relationship between matter and curvature in the Universe. Instead of adding a new matter component to the RHS of Einstein's equation, an alternative method to explain the accelerating expansion is to modify general relativity on the LHS of the equation. There is an abundance of modified gravity models on the market which can be motivated either by low energy limits of string theory, which generally feature a new scalar field degree of freedom, or by higher dimensional gravity theories, which change the dimensionality of space. We will briefly discuss a couple of examples (for a more detailed review see e.g. Jain and Khoury 2010).

In $f(R)$ gravity, the Einstein–Hilbert action is modified by the addition of a general function of the Ricci scalar,

$$S \sim \int d^4x \sqrt{-g}\, R \rightarrow \int d^4x \sqrt{-g}[R + f(R)]. \qquad (1.19)$$

In the absence of a cosmological constant, this $f(R)$ term induces a late time accelerating expansion. A simple example of one of these theories is $f(R) \propto 1/R$

(Carroll et al. 2004). These models are severely constrained by Solar System tests of general relativity (Hu and Sawicki 2007). In addition to these constraints there are several limits on the functional form of $f(R)$, for example, $1 + f(R) > 0$ for all R so that the effective gravitational constant $G_{\text{eff}} = G/(1 + f(R))$ is positive (Amendola et al. 2007). Scalar–tensor theories, first introduced by Brans and Dicke (1961), feature a scalar field in the Einstein–Hilbert action which is non-minimally coupled to the matter fields in the so called Einstein frame. $f(R)$ theories are formally equivalent to scalar–tensor theories where the two are related by a conformal transformation of the metric.[1] One example is 'extended quintessence' which can be understood as a theory of gravity with an effective Newton's constant which depends on the scalar field. The action for extended quintessence in the Jordan frame can be written as

$$S = \int d^4 x \sqrt{-g} \left(\frac{1}{2\kappa} F(\varphi, R) - \frac{1}{2} \kappa(\varphi) \varphi^{;\mu} \varphi_{;\mu} \right.$$
$$\left. - V(\varphi) + L_{\text{m}} \right), \tag{1.20}$$

where the scalar field, φ, describes the gravitational interaction and has kinetic and potential energy $\kappa(\varphi)$ and $V(\varphi)$, respectively. Note here natural units are used where $c = 1$. The parameter $\kappa = 8\pi G_N$. The determinant of the background metric is denoted by g which is generally assumed to be a flat FRW cosmology and L_{m} is the usual matter Lagrangian. In standard quintessence models $F(\varphi, R)$ is just given by the Ricci scalar, R, and the gravitational action is identical to that in general relativity. We discuss these models in more detail in Chap. 5.

Extra dimensional modified gravity theories, such as braneworld cosmologies, describe our (3+1) D Universe as being embedded in a higher dimensional spacetime. For example, the DGP (Dvali et al. 2000) model features an infinite volume 5th dimension where the cosmic acceleration of the Universe arises from gravity confined to the 4D brane where it can be described as an effective scalar–tensor theory. At large distances there is a cross-over scale, r_c, from the usual general relativity force law, $1/r^2$, to a corresponding 5D gravity, $\sim 1/r^3$. In a homogeneous and isotropic Universe the DGP cosmology allows two solutions where the cross over scale appears in the modified Friedmann equation as

$$H^2 \pm \frac{H}{r_c} = \frac{8\pi G \rho}{3}. \tag{1.21}$$

The self accelerating branch of DGP exhibits accelerated expansion at late times without including dynamical scalar fields or a cosmological constant and corresponds to choosing the minus sign in the above equation. The accelerating expansion arises from the effective weakening of gravity on large scales and in effect it can be described

---

[1] In the Einstein frame, the gravitational action is the same as in general relativity and the scalar field appears in the matter action. By re-scaling the metric, one can express the action in the so-called Jordan frame where the Einstein–Hilbert action of general relativity is modified by the introduction of a scalar field.

by a smaller gravitational constant on these scales. This cosmology has been shown to contain a number of pathologies, for example, in linear perturbation theory the self accelerating branch contains 'ghosts', kinetic terms with the wrong sign which suggests that the theory is unstable (Gregory et al. 2007). Recently the growth of dark matter perturbations in the normal branch of the DGP model (plus sign in Eq. 1.21), together with a cosmological constant, has been studied in N-body simulations (Schmidt et al. 2010).

1.4 Testing the Concordance Cosmological Model

Constraining the properties of dark energy and modified gravity models in future surveys will require precise measurements of the expansion history and the growth rate of structure using a number of observations, such as the CMB, supernovae light curves and the BAO already discussed. In this section we review other cosmological probes relevant for current and future surveys.

The Integrated Sachs Wolfe (ISW) effect in the cosmic microwave background arises due to time varying gravitational potentials which cause a differential redshift in CMB photon energies. These photons gain energy as they fall into the potential wells and lose it as they exit. At recent times, these potential wells are decaying due to the presence of dark energy and so there is an overall gain in the photon's energy as it traverses the potential. This leads to a boost in the large angle (low multipole) correlation amplitude in the CMB power spectrum. Although this large scale observation is limited by cosmic variance, the ISW effect has been measured by cross correlating CMB measurements with galaxy catalogues to identify non-primordial CMB signals (Pietrobon et al. 2006; Cabré et al. 2006; Giannantonio et al. 2008). The amount of dark energy and its clustering properties can also be tested by combining measurements of the ISW effect with other probes of the gravitational potentials such as weak lensing.

When light from a distant galaxy travels through intervening large scale structure on its way to us, the gravitational potential distorts the path of the light ray, causing the galaxy's image to be gravitationally lensed. The distortion of the image is referred to as 'shear' and is sensitive to both the expansion history of the Universe and the effect of dark energy on the gravitational potentials, Ψ and Φ. In the conformal Newtonian gauge Ψ and Φ represent scalar perturbations to the time and space components of the metric (see e.g. Ma and Bertschinger 1995) and are equal to one another in general relativity. Measurements of weak lensing shear allow us to map out the dark matter distribution in the Universe and its evolution in time, which will be affected by the late time accelerating expansion. The shear angular power spectrum is sensitive to both the geometry of the Universe, through the angular diameter distance and the weight function which describes the efficiency for lensing a population of galaxies, and the growth of structure through the matter power spectrum.

Clusters of galaxies represent the largest virialised structures in the Universe and can be used to probe the properties of dark energy by comparing the observed number of clusters in a given volume element with predictions from a dark energy model with a given expansion history and growth rate. Using N-body simulations we can measure the number density of cluster sized haloes of mass M, $dn(z)/dM$, at a certain redshift, z, as well as the volume element at that redshift, dV/dz, to obtain dn/dz in a given cosmology and compare with results from large area surveys which associate cluster observables such as X-ray temperature or luminosity with cluster mass (see e.g. White et al. 1993). In galaxy clusters most of the baryons are in the intervening gas and measurements of the baryon to total mass density fraction, $f_{gas} = \Omega_b/\Omega_m$, can be used to determine the cluster mass. Other observables such as the Sunyaev–Zel'dovich effect, where CMB photons are energised by hot cluster gas resulting in a decrease in the CMB intensity at low frequencies and an increase at high frequencies, or weak lensing shear can also be used to measure the cluster mass.

Measurements of the expansion history alone can tell us if the dark energy equation of state is $w = -1$ or if it evolves in time but they do not test the law of gravity. The rate at which cosmic structures grow is set by a competition between gravitational instability and the rate of expansion of the Universe. As a result combined measurements of the growth rate and the expansion history allow us to test the framework of general relativity. The growth of structure can be measured by analysing the distortions in the galaxy clustering pattern, when viewed in redshift space (i.e. when a galaxy's redshift is used to infer its radial position). Proof of concept of this approach at $z > 0$ came recently from Guzzo et al. (2008), see Fig. 1.5, who used spectroscopic data for 10,000 galaxies from the VIMOS-VLT Deep Survey (Le Fevre et al. 2005) to measure the growth rate of structure at redshift $z = 0.77$ to an accuracy of $\sim 40\%$ (see also Peacock et al. 2001). We discuss redshift space distortions in more detail in Chaps. 4 and 5. As can be seen in Fig. 1.5, to distinguish between competing explanations for the accelerating expansion of the Universe, we need to measure the growth of structure to an accuracy of a few percent over a wide redshift interval.

1.5 Current and Future Observational Probes

At present numerous projects and surveys are either underway or being proposed to discover the underlying cause of the accelerating expansion. All of these projects make use of one or more of the observational probes we have discussed above. Here we highlight a few ground-based and space-based surveys.

The Panoramic Survey Telescope and Rapid Response System (Pan-STARRS)-1 (Kaiser and Pan-STARRS Project Team 2005) is a wide area survey which is now operational on Mount Haleakala, surveying 30,000 deg^2 with standard photometric g-, r-, i-, z- and y-band filters. A planned ultra-deep field survey of 1200 deg^2 (PS-4), which would make use of 4×1.8 m telescopes, will be able to measure supernovae light curves, galaxy clustering and weak lensing and could be used to measure BAO. Because of the large redshift error when using photometric redshift estimates

Fig. 1.5 The growth rate as a function of redshift from Guzzo et al. (2008) with the measurement at $z = 0.77$ from the VVDS-Wide survey (*filled circle*) together with the predictions from various theoretical models as labelled in the key. The small error bars show an estimate of the level of error expected from Euclid

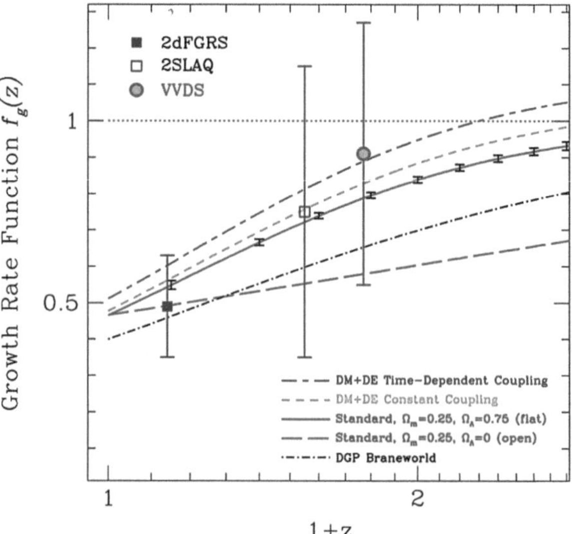

compared to spectroscopic ones, the appearance of the BAO may be damped and the number of useful modes in the measured power spectrum will be reduced, limiting the statistical power of such a measurement. Nevertheless, the volume covered by the main 3π survey of PS1 and the number of galaxies mapped make it worth investigating the measurement of the BAO feature in this survey. The WiggleZ Dark Energy Survey (Drinkwater et al. 2010; Blake et al. 2010), which began in 2006, is a large spectroscopic survey which aims to obtain 200,000 redshifts for UV-selected galaxies using the 3.9m Anglo-Australian Telescope. As of May 2010 the survey has obtained a total sample of 152,117 galaxy redshifts. The primary aim is to measure the BAO in the galaxy power spectrum, constraining the expansion history to better than 2 % and the growth rate to better than 20 % from redshift space distortions in the redshift range $0.2 < z < 1$. The SDSS-III's Baryon Oscillation Spectroscopic Survey (BOSS) (Schlegel et al. 2007) currently operating in New Mexico is a galaxy redshift survey of 1.5 million luminous red galaxies (LRGs) at $0.2 < z < 0.8$. BOSS will map out the BAO signal and obtain absolute distance measurements to a precision of 1 % at $z < 0.6$ with a sky coverage of 10,000 deg^2. The Large Synoptic Survey Telescope (LSST) (Ivezic et al. 2008) is an ambitious future project which will become operational by 2018 which will use a 8.4m ground based telescope in Northern Chile. The deep-wide-fast survey mode will cover a 20,000 deg^2 region over 10 years of operations measuring multiple probes of dark energy, most notably BAO and weak gravitational lensing tomography. The Hobby-Eberly Telescope Dark Energy EXperiment (HETDEX) (Hill et al. 2004) will measure the BAO using the redshifts of millions of Ly-α emitting galaxies in the redshift range $2 < z < 4$. The aim is to constrain the expansion history and the growth rate out to $z = 2.4$ to 0.8 % and 2 % respectively. BigBOSS is a proposed ground based spectroscopic survey which will measure the expansion history and the growth rate to sub-percent level accuracy over redshifts $0 < z < 3.5$ looking at BAO and redshift space distortions in the galaxy power spectrum. The Wide-Field Multi-Object Spectrograph (WFMOS) (Bassett et al. 2005) is a proposed project with the Subaru 8.2m telescope which will measure BAO in the galaxy power spectrum at $z < 1.3$.

The European Space Agency (ESA) currently has one funded dark energy mission, eROSITA, and another dark energy mission, Euclid, under consideration. eROSITA (extended ROentgen Survey with an Imaging Telescope Array) is a German–French collaboration which aims to detect 50–100 thousand clusters of galaxies at $z \sim 1.3$ (Predehl et al. 2006). The second misson, Euclid (Cimatti et al. 2009), has emerged from combining the Dark Universe Explorer (DUNE) and the SPACE concepts which aim to measure weak lensing and baryonic acoustic oscillations at redshifts $0.5 < z < 2$. In Chap. 5 we measure the redshift space distortions in a N-body simulation of a modified gravity cosmology to test the accuracy of current models for the redshift space power spectrum in recovering the correct value for the growth rate at $z = 0.5$. Our simulation volume of $1500h^{-1}$Mpc cubed corresponds to a similar comoving volume available to Euclid at $z = 0.5$ assuming a sky coverage of 20,000 deg^2 and a redshift shell of thickness $\Delta z = 0.1$.

Based on several space based missions considered for the NASA-DOE Joint Dark Energy mission (JDEM) (Gehrels 2010), the Wide Field Infrared Survey Telescope (WFIRST) has been proposed in the US decadal review 'New Worlds, New Horizons in Astronomy and Astrophysics' (Gould 2010). WFIRST is a 1.5m infrared telescope which plans to image about 2 billion galaxies in order to study weak lensing, probing both the expansion rate and the growth of structure. WFIRST also aims to measure BAO by obtaining the spectra of about 200 million galaxies and will be able to detect thousands of supernovae providing two robust measurements of the expansion history.

The huge investment of human resources and funding dedicated to probing the properties of dark energy and modified gravity in future surveys is clear and needs to be matched by precise predictions and models calibrated using N-body simulations. Accurate modelling of the linear, quasi-linear and non-linear regimes is essential for interpreting future surveys whose total volume will reach $\sim$10s of Gpc^3. This thesis focuses on measuring several key observational probes of dark energy and general relativity from consistent N-body simulations of different cosmologies, namely the clustering of matter on large scales, the halo mass function, baryonic acoustic oscillations and redshift space distortions. These results can be used to extend the statistical power of future galaxy surveys.

1.6 Outline of Thesis

The growth of large scale structure in the Universe is an extremely important tool which can be used to probe fundamental physics such as the nature of dark energy or modified gravity theories. Cosmological N-body simulations play a vital role in cosmology for both theorists and observers and are an important laboratory where we can test current theories of structure formation. The results presented in this work represent a step forward in simulating quintessence dark energy models in ultra-large volume computational boxes. With many precision tests of dark energy and modified gravity planned in future galaxy surveys, the aim of this work is to improve the current models and predictions for observables using accurate simulations of alternative cosmologies. Using the N-body simulations presented here we can answer some of the key questions posed by future surveys, such as, can we detect a variation in $w(z)$ by measuring BAO peak positions to within $1\,\%$? or can we distinguish a modification to gravity from dark energy with a measurement of the growth rate which is accurate to $\sim 2\,\%$?

The main goals of this thesis can be summarised as follows: firstly in Chap. 3 we consider viable quintessence dark energy cosmologies and conduct consistent N-body simulations of these models, fully accounting for the different expansion histories, modified linear theory and different values of the cosmological parameters which are needed to match current observations. We study the non-linear growth of cosmic structure in these models and compare the growth of structure to that in a universe with a cosmological constant. Using these N-body simulations we measure the non-linear power spectra, the halo mass function, the BAO peak positions and the redshift space distortions in different quintessence dark energy models and test for detectable differences from the standard ΛCDM model. In Chap. 4 we focus on the use of redshift space distortions as a probe of the growth rate of structure which has been suggested as a key observable with which to test general relativity. In Chap. 5 we conduct N-body simulations of two competing cosmologies—a dark energy model with a scalar field and the other with a change to Newton's gravitational constant. We test the accuracy of several models for the redshift space distortions and their ability to recover the correct growth factor which would distinguish modified gravity from dark energy. A summary of the thesis is presented in Chap. 6.

References

Aguirre AN (1999) ApJ 512:L19
Amendola L, Gannouji R, Polarski D, Tsujikawa S (2007) Phys Rev D 75:083504
Angulo R, Baugh CM, Frenk CS, Lacey CG (2008) MNRAS 383:755
Astier P et al (2005) Bull Am Astron Soc 37:1176
Bassett BA, Nichol B, Eisenstein DJ (2005) Astron Geophys 46:050000
Bennett CL, Banday AJ, Gorski KM, Hinshaw G, Jackson P, Keegstra P, Kogut A, Smoot GF, Wilkinson DT, Wright EL (1996) ApJ 464:L1+
Bianchi E, Rovelli C (2010) ArXiv e-prints
Blake C, Glazebrook K (2003) ApJ 594:665
Blake C et al (2010) MNRAS 406:803
Brans C, Dicke RH (1961) Phys Rev 124:925
Cabré A, Gazta naga E, Manera M, Fosalba P, Castander F (2006) MNRAS 372:L23
Caldwell RR, Dave R, Steinhardt PJ (1998) PhRvL 80:1582
Carroll SM (2001) Living Rev Rel 4:1

Carroll SM, Duvvuri V, Trodden M, Turner MS (2004) Phys Rev D 70:043528

Cimatti A et al (2009) Exp Astron 23:39

Cole S et al (2005) MNRAS 362:505

Copeland EJ, Sami M, Tsujikawa S (2006) Int J Mod Phys D 15:1753

Drinkwater MJ et al (2010) MNRAS 401:1429

Dvali GR, Gabadadze G, Porrati M (2000) Phys Lett B 485:208

Efstathiou G et al (2002) MNRAS 330:L29

Eisenstein DJ et al (2005) ApJ 633:560

Ferreira PG, Joyce M (1998) PhysRev D 58:023503

Le Fevre O et al (2005) A&A 439:845

Gehrels N (2010) Status of the Joint Dark Energy Mission (JDEM). In: Bulletin of the American astronomical society, vol 42, p 590

Giannantonio T, Scranton R, Crittenden RG, Nichol RC, Boughn SP, Myers AD, Richards GT (2008) Phys Rev D 77:123520

Gould A (2010) ArXiv e-prints

Gregory R, Kaloper N, Myers RC, Padilla A (2007) J High Energy Phys 10:69

Guzzo L et al (2008) Nature 451:541

Hill GJ, Gebhardt K, Komatsu E, MacQueen PJ (2004) The Hobby-Eberly Telescope Dark Energy Experiment. In: Allen RE, Nanopoulos DV, Pope CN (ed) The new cosmology: conference on strings and cosmology, American Institute of Physics Conference Series, vol 743, pp 224–233

Hu W, Haiman Z (2003) Phys Rev D 68:063004

Hu W, Sawicki I (2007) Phys Rev D 76:064004

Ivezic Z, Tyson JA, Allsman R, Andrew J, Angel R (2008) LSST Collaboration 2008, ArXiv e-prints

Jain B, Khoury J (2010) Ann Phys 325:1479

Kaiser N (2005) Pan-STARRS Project Team 2005, The Pan-STARRS large survey telescope project. In: Bulletin of the American astronomical society, vol 37, p 622

Knop RA et al (2003) ApJ 598:102

Komatsu E et al (2010) ArXiv e-prints

Kowalski M et al (2008) ApJ 686:749

Larson D et al (2010) ArXiv e-prints

Linder EV (2008) Gen Rel Grav 40:329

Ma C, Bertschinger E (1995) ApJ 455:7

Meiksin A, White M, Peacock JA (2000) Nucl Phys B Proc Suppl 80:C913+

Miknaitis G et al (2007) ApJ 666:674

Oyaizu H (2008) Phys Rev 78:123523

Peacock JA et al (2001) Nature 410:169

Perlmutter S et al (1999) ApJ 517:565

Phillips MM (1993) ApJ 413:L105

Pietrobon D, Balbi A, Marinucci D (2006) Phys Rev D 74:043524

Predehl P et al (2006) In: Society of photo-optical instrumentation engineers (SPIE) conference series, vol 6266, eROSITA

Ratra B, Peebles PJE (1988) Phys Rev D 37:3406

Riess AG et al (1998) ApJ 116:1009

Riess AG et al (2004) ApJ 607:665

Sánchez AG, Crocce M, Cabré A, Baugh CM, Gazta naga E (2009) MNRAS 400:1643

Sanchez AG et al (2006) MNRAS 366:189

Schlegel DJ et al (2007) The Baryon Oscillation Spectroscopic Survey (BOSS). In: Bulletin of the American astronomical society, SDSS-III, vol 38, p 966

Schmidt F, Hu W, Lima M (2010) Phys Rev D 81:063005

Wetterich C (1988) Nucl Phys B 302:668

White SDM, Navarro JF, Evrard AE, Frenk CS (1993) Nature 366:429

Wood-Vasey WM et al (2007) ApJ 666:694

Chapter 2
The Growth of Matter Perturbations in the Universe

2.1 Numerical Methods

In this chapter we outline some aspects of the N-body simulation code used in this thesis as well as the modifications made to the code to include the effects of various dark energy cosmologies. We also describe how the initial conditions for the simulations are set up.

2.1.1 The Simulation Code

Once a dark matter perturbation approaches the cosmic mean, $\delta \sim 1$, linear theory breaks down and full numerical methods are needed in order to follow the non linear growth of structure. Analytic solutions can be used in special circumstances, for example, the Press-Schechter formalism can be used to predict the number of objects of a certain mass in a given volume assuming spherical collapse (Press and Schechter 1974). Here we present a brief review of the N-body simulation code GADGET- 2. For more information on the code see Springel (2005) and for a comprehensive review of N-body simulations see Bertschinger (1998).

Following the dynamics of dark matter particles under their mutual gravitational attraction requires us to solve the collisionless Boltzmann equation and Poisson's equation simultaneously. Using a method of characteristics (Leeuwin et al. 1993) the solution of the Boltzmann equation can be obtained by sampling the $(6 +1)$ dimensional phase space, $\{\vec{x}, \vec{p}, t\}$, of the initial distribution function, $f(\vec{x}, \vec{p}, t)$. Solving Poisson's equation for N particles, the system can be evolved forward in time using the equations of motion derived from $\partial f/\partial t + [f, H] = 0$, where H, in this instance, is the system's Hamiltonian.

The core of any N-body simulation is the gravity solver. In the PM (particle-mesh) algorithm the density field is realised on a grid and the gravitational potential is constructed by solving Poisson's equation. In this scheme all the particles are

E. Jennings, *Simulations of Dark Energy Cosmologies*, Springer Theses,
DOI: 10.1007/978-3-642-29339-9_2, © Springer-Verlag Berlin Heidelberg 2012

assigned to a grid using a kernel which splits up the masses and determines the density field, $\rho_{i,j,k}$, at each grid point. The simplest choice of mass assignment scheme is nearest grid point (NGP) where all the mass is allocated to the nearest grid cell. This method leads to significant fluctuations in the evaluated force which can be avoided by using higher order schemes such as the cloud-in-cell (CIC) or triangular shaped cloud (TSC) schemes (Hockney and Eastwood 1981). In the CIC scheme the mass is assigned to the 8 grid points nearest to the particle while the TSC method uses the nearest 27 grid points. The kernel used to construct the density field in the PM part of GADGET- 2 is the CIC assignment scheme. The density field on the grid is then Fourier transformed and the potential on the grid is obtained using the Green's function, $-4\pi G/k^2$, to solve Poisson's equation, $\nabla^2 \phi_{i,j,k} = 4\pi G\rho_{i,j,k}$ in Fourier space. Using a grid to estimate the forces in this way results in a lack of short range accuracy on scales comparable to the grid spacing. The Particle-Particle PM scheme (P^3M) overcomes the force resolution problem associated with PM methods by adding a direct summation of pairs separated by less than 2 or 3 grid spacings. The combination of mesh based and direct pair summation results in high accuracy forces. However, the algorithm slows down when clustering becomes strong on small scales which degrades the performance of the P^3M code.

GADGET- 2 makes use of a TreePM algorithm to compute the gravitational forces accurately. The tree algorithm groups distant particles into larger cells and approximates their potentials using multipole expansions about the centre of mass of the group (Barnes and Hut 1986). The advantage of this method is a scaling in computation time of $\mathcal{O}(N\log N)$, where N is the number of particles, compared to $\mathcal{O}(N^2)$ calculations with a direct summation of the forces. The error on the long range force is then controlled by an opening angle parameter which determines when a multipole expression is used to calculate the forces for a group of particles. A distant cell of mass M, at a distance r and extension l, is considered for opening if

$$\frac{GM}{r^2}\left(\frac{l}{r}\right)^2 \leq \alpha|a| \tag{2.1}$$

where α is a tolerance parameter and a is the total acceleration obtained in the last timestep. The TreePM algorithm employed in GADGET- 2 combines the computational efficiency of the PM code with the short range accuracy of the tree code and splits the gravitational potential into a long and short range component, $\Phi = \phi^{\text{short}} + \phi^{\text{long}}$, where the tree algorithm is used to evaluate the force on small scales and the long range potential is calculated using a mesh. The spatial scale of the force split, r_s, is present in the expression for the short range potential given by

$$\phi^{\text{short}}(\vec{x}) = -G\sum_i \frac{m_i}{r_i}\text{erfc}\left(\frac{r_i}{2r_s}\right), \tag{2.2}$$

where the smallest distance of any of the images of a particle, i, in a periodic box of length L, to the point $\vec{x}$ is given by $r_i = \min[|\vec{x} - \vec{r}_i - \vec{n}L|]$. The force is estimated according to $F_{i,j,k} = -\nabla \Phi_{i,j,k}$ by finite differencing the potential. The force is then interpolated back to the particle positions using the CIC kernel.

To avoid a singularity in the force calculation when particle separations are close to zero, it is common to introduce a softening parameter which softens the force and limits the maximum relative velocity during close encounters between particles. This softening also prevents the artificial formation of binaries in the simulation. The equations of motion in an expanding Universe are obtained by integrating Hamilton's equations

$$\frac{d\vec{x}}{dt} = \frac{\vec{p}}{a^2}, \tag{2.3}$$

$$\frac{d\vec{p}}{dt} = -\frac{\nabla \Phi}{a}, \tag{2.4}$$

where $\vec{p} = a^2 m \vec{x}$ is the canonical momentum and Φ is the interaction potential. In GADGET-2 these equations are discretized and integrated using 'kick' and 'drift' operators in a second order accurate leap frog integrator scheme (Springel 2005). The drift and kick operators are the time evolution operators of the kinetic and potential components of the Hamiltonian of the N-body problem. The drift operator leaves the momentum unchanged and advances the position of each particle, while the kick operator leaves the position unchanged and updates the momentum. In one time step a combination of these is used, for example the drift-kick-drift (DKD) leapfrog integrator. For each particle the timestep in GADGET-2 is given by

$$\Delta t = \min \left[\Delta t_{\text{max}}, \left(\frac{2\eta\varepsilon}{a} \right)^{1/2} \right], \tag{2.5}$$

where ε is the gravitational softening, η is an accuracy parameter, a is the particle's acceleration and Δt_{max} can be set to a fraction of the dynamical time of the system. We discuss the initial conditions of the N-body code in Sect. 2.1.3.

2.1.2 Modifying Gadget-2

In this thesis we will determine the impact of quintessence dark energy on the growth of cosmological structures through a series of large N-body simulations. These simulations were carried out at the Institute of Computational Cosmology using a memory efficient version of the TreePM code `Gadget-2`, called `L-Gadget-2` (Springel 2005). As our starting point, we consider a ΛCDM model with the following cosmological parameters: $\Omega_m = 0.26$, $\Omega_{DE} = 0.74$, $\Omega_b = 0.044$, $h = 0.715$ and a spectral index of $n_S = 0.96$ (Sánchez et al. 2009). The linear theory rms fluctuation in spheres of radius $8\,h^{-1}$ Mpc is set to be $\sigma_8 = 0.8$.

Within the code of L-Gadget-2, under the assumption that the dark energy is a smooth background, the only place where dark energy needs to be accounted for within the code of L-Gadget-2, is in the calculation of the Hubble factor. This is needed, for example, when converting from the internal time variable, $\log a$ to a physical time, t, or when converting to physical quantities in the equations of motion. The Hubble parameter for dynamical dark energy in a flat universe is given by

$$\frac{H^2(z)}{H_0^2} = \left(\Omega_m \, (1+z)^3 + (1 - \Omega_m) e^{3 \int_0^z \mathrm{dln}(1+z')\,[1+w(z')]} \right), \qquad (2.6)$$

where H_0 and $\Omega_m = \rho_m/\rho_{crit}$ are the values of the Hubble parameter and dimensionless matter density, respectively, at redshift $z = 0$ and $\rho_{crit} = 3H_0^2/(8\pi G)$ is the critical density. The details of the dark energy equation of state, $w(z)$, for each quintessence model are given in Chap. 3.

In Chaps. 3 and 4, the simulations use $N = 646^3 \sim 269 \times 10^6$ particles to represent the dark matter in a computational box of comoving length $1{,}500\,h^{-1}$ Mpc. These simulations took 3 days to run with typically ~ 3000 time steps on 38 processors of the Cosmology Machine (COSMA) at Durham university. We chose a comoving softening length of $\varepsilon = 50\,h^{-1}$ kpc. The particle mass in the simulation is $9.02 \times 10^{11}\,h^{-1}M_\odot$ with a mean interparticle separation of $r \sim 2.3\,h^{-1}$ Mpc. The simulation code L-Gadget-2 has an inbuilt friends-of-friends (FOF) group finder which was applied to produce group catalogues of dark matter particles with 10 or more particles. A linking length of 0.2 times the mean interparticle separation was used in the group finder (Davis et al. 1985).

In Chap. 5 the simulations use $N = 1024^3 \sim 1 \times 10^9$ particles in a computational box of comoving length $1{,}500\,h^{-1}$ Mpc. The comoving softening length was $\varepsilon = 50\,h^{-1}$ kpc and the simulations took 5 days to run on 128 processors on COSMA. The L-Gadget-2 simulation code (Springel 2005) was modified to allow for a time-varying Newton's constant and a dynamical quintessence dark energy. As discussed in the previous section, in this code the gravitational forces are computed using a TreePM algorithm where short-range forces are calculated using a 'tree' method and the long-range part of the force is obtained using mesh based Fourier methods. In the modified gravity simulation, both the long and short-range force computations were modified to include a time-dependent gravitational constant. For both the modified gravity and the quintessence dark energy simulations in Chap. 5 the Hubble parameter computed by the code was also modified (see Chap. 5 for details).

2.1.3 The Initial Conditions

There are two steps needed to set up the initial conditions for an N-body simulation. In the first step an unperturbed Universe is created by setting up a uniform distribution of particles which, in the second step, is perturbed so that the resulting density

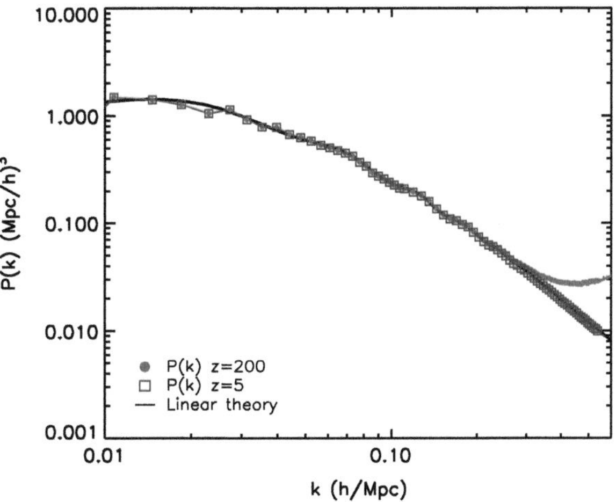

Fig. 2.1 The power spectrum measured from the simulation at $z = 200$ (*circles*) together with the power spectrum at $z = 5$ (*squares*) scaled to $z = 200$ by the squared ratio of the growth rates at the two redshifts. The linear perturbation theory prediction is shown as a *black line*

distribution has the appropriate power spectrum. An initially random distribution of particles will evolve into rapidly growing non linear structures due to the presence of Poisson shot noise on all scales. The initial 'white noise spectrum', in this case, is $|\delta_k|^2 \propto k^n$ where $n = 0$. A better way to generate a uniform distribution is to place the particles on a regular cubic grid, where there is no power above the nyquist frequency of the grid. This method also has its disadvantages as the regularity and size of the grid is imprinted as a characteristic length scale which is visible in the evolved particle distribution. Another method used to generate a uniform distribution of particles which has no regular structure, involves firstly placing the particles at random in a simulation volume. An N-body simulation code, which has been modified by reversing the sign of the acceleration, then follows the motion of the particles in an Einstein de Sitter expanding Universe. As a result the gravitational forces on the particles are repulsive and after many expansion factors they settle down to a 'glass-like' configuration where the distribution is sub-random and shows no order or anisotropy on scales comparable to the mean interparticle spacing (White 1994a; Baugh et al. 1995). The initial conditions of the particle load for the simulations in this thesis were set up with a glass configuration of particles.

In order to impose the density perturbations on the glass, the particles are perturbed using the Zel'dovich approximation (Zel'Dovich 1970) which moves the initially unperturbed particles to create a discrete density field using

$$\vec{x} = \vec{x}_0 - \frac{D(\tau)}{4\pi G \bar{\rho} a^3} \nabla \Phi_0 \tag{2.7}$$

$$\vec{v} = -\frac{1}{4\pi G \bar{\rho} a^2} \frac{a \dot{D}}{D} \nabla \Phi, \tag{2.8}$$

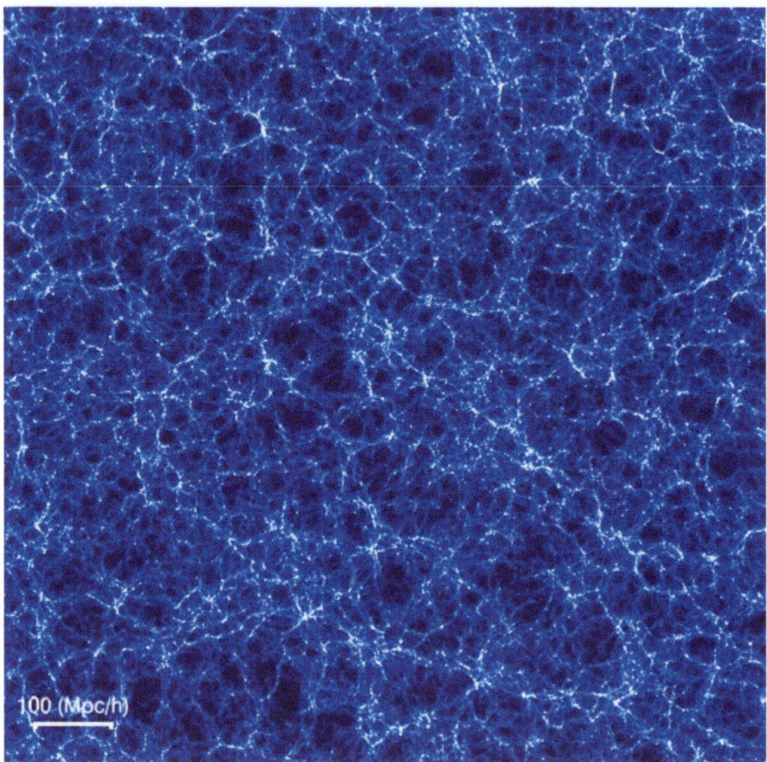

Fig. 2.2 The *dark* matter distribution from a simulation using 646^3 particles to represent the dark matter distribution in box of $1,500\,h^{-1}$ Mpc on a side at redshift $z = 0$

where the Eulerian position, $\vec{x}$, and the peculiar velocity, $\vec{v}$, of each particle are given as a function of its initial Lagrangian position, $\vec{x}_0$, and $D(\tau)$ is the growing mode of linear fluctuations as a function of conformal time, $d\tau = a^{-1}dt$ (e.g. Efstathiou et al. 1985; White 1994b). The displacement field, $\nabla\Phi$, is related to a precalculated input power spectrum, $P(k)$, with the desired cosmology. The initially uniform density field is then realised as a Gaussian random field with a random phase. The Zel'dovich approximation can induce small scale transients in the measured power spectrum. These transients die away after $\simeq 10$ expansion factors from the starting redshift (Smith et al. 2003). In order to limit the effects of the initial displacement scheme we chose a starting redshift of $z = 200$. In this thesis the linear theory power spectrum used to generate the initial conditions was created using the CAMB package of Lewis and Bridle (2002). The linear theory $P(k)$ output at $z = 0$ was then evolved backwards to the starting redshift of $z = 200$ using the linear growth factor for that cosmology in order to generate the initial conditions for L-Gadget-2. The details of the linear power spectra used for each dark energy model is outlined in Chap. 3. The initial power spectrum output at $z = 200$ is shown in Fig. 2.1 (circles) together with the linear perturbation theory (black line) and the power spectrum output at $z = 5$ (squares) scaled to $z = 200$ using the squared ratio of the growth rates at the two redshifts. The power spectrum is drawn from a distribution which results in fluctuations at low k, on large scales, due to the finite number of modes available in the simulation volume. The sample variance fluctuation can be clearly seen in the $z = 200$ and the $z = 5$ power spectra on large scales. The $z = 5$ output agrees very well with linear perturbation theory. In subsequent chapters in this thesis we shall use the $z = 5$ output in ratios to show deviations in growth from linear theory and to remove the sample variance present on large scales. In Fig. 2.2 we plot the dark matter distribution at $z = 0$ in a 2D slice through the simulation with 646^3 particles in a box of $1{,}500\,h^{-1}$ Mpc in length.

References

Barnes J, Hut P (1986) Nature 324:446
Baugh CM, Gaztanaga E, Efstathiou G (1995) MNRAS 274:1049
Bertschinger E (1998) ARA&A 36:599
Davis M, Efstathiou G, Frenk CS, White SDM (1985) ApJ 292:371
Efstathiou G, Davis M, White SDM, Frenk CS (1985) ApJS 57:241
Hockney RW, Eastwood JW (1981) Computer simulation using particles. McGraw-Hill, New York
Leeuwin F, Combes F, Binney J (1993) MNRAS 262:1013
Lewis A, Bridle S (2002) Phys. Rev. D 66:103511
Press WH, Schechter P (1974) ApJ 187:425
Sánchez AG, Crocce M, Cabré A, Baugh CM, Gaztañaga E (2009) MNRAS 400:1643
Smith RE et al (2003) MNRAS 341:1311
Springel V (2005) MNRAS 364:1105
White SDM (1994) ArXiv Astrophysics e-prints
White SDM (1994) RvMA 7:255
Zel'Dovich YB (1970) A&A 5:84

Chapter 3
Simulations of Quintessential Cold Dark Matter

3.1 Introduction

Quintessence models of dark energy are studied as a viable alternative to the cosmological constant and feature an evolving scalar field which dominates the energy budget today causing accelerated expansion. In this chapter we present three stages of N-body simulations of structure formation in quintessence models. Each stage progressively relaxes the assumptions made and brings us closer to a full physical model. In the first stage, the initial conditions for each quintessence cosmology are generated using a ΛCDM linear theory power spectrum and the background cosmological parameters are the best fit values assuming a ΛCDM cosmology. The only departure from ΛCDM in this first stage is the dark energy equation of state and its impact on the expansion rate. In the second stage, we use a modified version of CAMB (Lewis and Bridle 2002) to generate a consistent linear theory power spectrum for each quintessence model. The linear theory power spectrum can differ from the power spectrum in ΛCDM due to the presence of non-negligible amounts of dark energy during the early stages of the matter dominated era. This power spectrum is then used to generate the initial conditions for the N-body simulation which is run again for each dark energy model. The third and final stage in our analysis is to find the values for the cosmological parameters, $\Omega_m h^2$, $\Omega_b h^2$ and H_0 (the matter density, baryon density and Hubble parameter) such that each model satisfies cosmological distance constraints. Recently Alimi et al. (2010) used CMB and SN data to constrain the parameters in the quintessence potential and the value of the matter density, $\Omega_m h^2$, for two models. In this chapter we allow three parameters to vary when fitting each quintessence model to the available data. This distinction is important as changes in these parameters may produce compensating effects which result in the quintessence model looking like ΛCDM. For example, for a given dark energy equation of state, a lower value of the matter density may not result in large changes in the Hubble parameter if the value of H_0 is increased. In going through each of these stages we build up a comprehensive picture of the quintessence models and their effect on the nonlinear growth of structure.

E. Jennings, *Simulations of Dark Energy Cosmologies*, Springer Theses,
DOI: 10.1007/978-3-642-29339-9_3, © Springer-Verlag Berlin Heidelberg 2012

This chapter is organised as follows. In Sect. 3.2 we discuss quintessence models and the parametrization we use for the dark energy equation of state. We also outline the expected impact of different dark energy models on structure formation. The main power spectrum results are presented in Sect. 3.3. Intermediate results are presented in Sects. 3.3.1 and 3.3.2. In Sect. 3.3.4 we present the mass function predictions. In Sect. 3.3.5 we discuss the appearance of the baryonic acoustic oscillations in the matter power spectrum. Finally, in Sect. 3.4 we present our summary.

3.2 Quintessence Models of Dark Energy

Two broad classes of quintessence models can be used to solve both the fine-tuning and coincidence problems. The first is based on the idea of so called 'tracker fields' (Steinhardt et al. 1999). These fields adapt their behaviour to the evolution of the scale factor and hence track the background density. The other class is referred to as 'scaling solutions' (Halliwell 1987; Wands et al. 1993; Wetterich 1995). In these models the ratio of energy densities, ρ_φ/ρ_B, is constant.

In tracking models, the φ field rolls down its potential, $V(\varphi)$, to an attractor-like solution. The great advantage of these models is that this solution is insensitive to the initial conditions of the scalar field produced after inflation. A general feature of these tracking solutions is that as the scalar field is tracking behind the dominant matter component in the universe, its equation of state, w_φ, depends on the background component as

$$\frac{\rho_\varphi}{\rho_B} = a^{3\,(w_B - w_\varphi)}, \tag{3.1}$$

where ρ_B and w_B denote the background energy density and equation of state respectively, with $w_B = 1/3$ (radiation era) and $w_B = 0$ (matter era). As a result, the energy density of the scalar field remains sub-dominant during the radiation and matter dominated epochs, although it decreases at a slower rate than the background density. The quintessence field, ρ_φ, naturally emerges as the dominant component today and its equation of state is driven towards $w = -1$. An example of a tracking model is the inverse potential form proposed by Zlatev et al. (1999), $V(\varphi) \sim M^{4+\alpha}\varphi^{-\alpha}$, where M is a free parameter that is generally fixed by the requirement that the dark energy density today $\Omega_{DE} \sim 0.7$ and so the quintessence potential must be $V \sim \rho_{crit}$. This implies that φ is of the order of the Planck mass today, $\varphi \sim M_{Pl}$. With $\alpha \leq 6$, the quintessence field equation of state is approximately $w_0 \lesssim -0.4$ today.

In scaling quintessence models, the ratio of energy densities, ρ_φ/ρ_B, is kept constant, unlike tracking models, where ρ_φ changes more slowly than ρ_B. During the evolution of the energy density in a 'scaling' model, if the dominant matter component advances as $\rho \propto a^{-n}$, then the scalar field will obey $\Omega_\varphi = n^2/\alpha^2$ after some initial transient behaviour. Scaling quintessence models can suffer from an inability to produce late time acceleration, whilst at the same time adhering to observational constraints, such as, for example, the lower limit on Ω_φ during nucleosynthesis

(Bean et al. 2001). Albrecht and Skordis (2000) used a modified coefficient in their scaling potential, $V(\varphi) = V_{\mathrm{p}}\, e^{-\lambda \varphi}$, where $V_{\mathrm{p}}(\varphi) = (\varphi - B)^{\alpha} + A$, resulting in a model which can produce late time acceleration as well as satisfying cosmological bounds, for a variety of constants A and B. Barreiro et al. (2000) considered a linear combination of exponential terms in the scalar field potential and found this yielded a larger range of acceptable initial energy densities for φ compared with inverse models. Copeland et al. (2000) also consider supergravity (SUGRA) corrections to quintessence models, where the resulting potential can exhibit either 'tracking' or 'scaling' behaviour depending on which path the scalar field takes down its potential towards the minimum where it would appear as a cosmological constant.

The physical origin of the quintessence field should be addressed by models motivated by high energy particle physics. As the vacuum expectation value of the scalar field today is of the order of the Planck mass, any candidates for quintessence which arise in supersymmetric (SUSY) gauge theories may receive supergravity corrections which will alter the field's potential. It is this fact that motivates many authors to argue that any quintessence model inspired by particle physics potentials must be based on SUGRA. Brax and Martin (1999) discuss such models and employ the potential $V(\varphi) = \Lambda^{4+\alpha}/\varphi^{\alpha} e^{\kappa/2\varphi^2}$ with a value of $\alpha \geq 11$ in order to drive w_0 close to -1 today.

In summary, in this chapter we will consider six quintessence models which cover the behaviours discussed above. In particular, INV1 and INV2, which are plotted in Fig. 3.1, have inverse power law potentials and exhibit tracking solutions. The INV1 model is the 'INV' model considered by Corasaniti and Copeland (2003) and has a value of $w_0 = -0.4$ today. As current observational data favour a value of $w_0 < -0.8$ (Sánchez et al. 2009), the INV1 model will be used as an illustrative model. We shall consider a second inverse power law model (INV2) which is in better agreement with the constraints on w. As noted by Corasaniti (2004), the scale Λ in the inverse power law potential, $V(\varphi) = \Lambda^{\alpha+4}/\varphi^{\alpha}$ is fixed by the value of Ω_{DE} today. Solving the coincidence problem requires this scale for Λ to be consistent with particle physics models. For values of $\alpha \geq 6$ it is possible to have energy scales of $\Lambda \sim 10^6$ GeV. Setting $\alpha = 6$ results in an equation of state with $w_0 = -0.4$ (INV1). It is possible to drive the equation of state closer to -1 today with lower values of α, although the value of Λ is then pushed to an undesirable energy range when compared with the typical scales of particle physics. The second model INV2, which has $w_0 = -0.79$ with $\alpha = 1$, has been added to illustrate a power law potential with a dark energy equation of state which agrees with constraints found on w_0 using CMB, SN and large scale structure data (Sánchez et al. 2009). We also use the SUGRA model of Brax and Martin (1999) which exhibits tracking field behaviour. The potential in this case also contains an exponential term which pushes the dark energy equation of state to $w_0 = -0.82$. The 2EXP model is an example of a scaling solution and features a double exponential term in the scalar field potential (Barreiro et al. 2000). The AS model suggested by Albrecht and Skordis (2000) belongs to the class of scaling quintessence fields. As mentioned previously, the parameters in this potential can be adjusted to have the fractional dark energy density, Ω_{DE}, below the nucleosynthesis

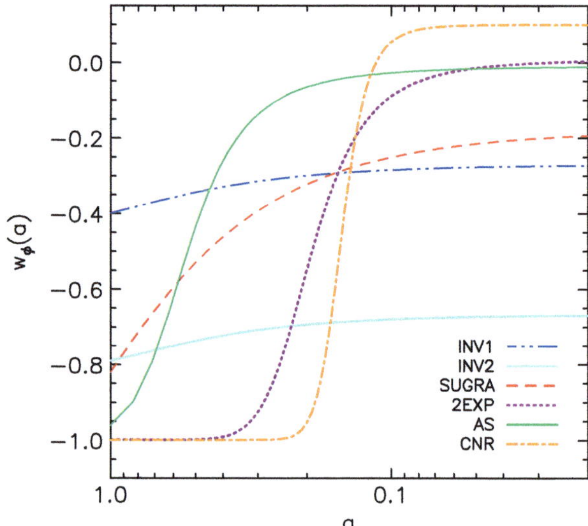

Fig. 3.1 The dark energy equation of state as a function of expansion factor, $w(a)$, for six quintessence models motivated by particle physics, which are either tracking or scaling solutions. The parametrization for $w(a)$ is given in Eq. 3.2 and the four parameter values which specify each model are given in Table 3.1. Note the *left hand side* of the x-axis is the present day

bound in the early universe. The CNR model (Copeland et al. 2000) has a tracking potential where the scalar field rolls down to its minimum and will settle down to $w_0 = -1$ after a series of small oscillations.

Each of the quintessence models we consider is one of a family of such models with parameter values chosen in order to solve the issues of fine-tuning and coincidence, as well as to produce a value of $w_0 \sim -1$ today. These requirements limit the parameter space available to a particular quintessence potential. For example, this limits the range of the Brax and Martin (1999) SUGRA model. The SUGRA model we simulate has a fixed parameter value in the supergravity potential but the dark energy equation of state for this model does not depend strongly on this parameter (see Fig. 4 in Brax and Martin 1999).

3.2.1 Parametrization of w

Given the wide range of quintessence models in the literature it would be a great advantage, when testing these models, to obtain one model independent equation describing the evolution of the dark energy equation of state without having to specify the potential $V(\varphi)$ directly. Throughout this chapter we will employ the parametrization for w proposed by Corasaniti and Copeland (2003), which is a generalisation of the method used by Bassett et al. (2002) for fitting dark energy models with rapid late time transitions. Using a parametrization for the dark energy equation of state provides us with a model independent probe of several dark energy properties. The dark energy equation of state, $w(a)$, is described by its value during radiation domination, w_r, followed by a transition to a plateau in the matter dominated era, w_m, before making the transition to the present day value w_0. Each of these transitions can be parametrized by the scale factor $a_{r,m}$ at which they occur and the width of the transition $\Delta_{r,m}$.

In order to reduce this parameter space we use the shorter version of this parametrization for w, which is relevant as our simulations begin in the matter dominated era. The equation for w valid after matter-radiation equality is

$$w_\varphi(a) = w_0 + (w_m - w_0) \times \frac{1 + e^{\frac{a_m}{\Delta_m}}}{1 + e^{-\frac{a - a_m}{\Delta_m}}} \times \frac{1 - e^{-\frac{a-1}{\Delta_m}}}{1 - e^{\frac{1}{\Delta_m}}} . \tag{3.2}$$

Corasaniti and Copeland (2003) showed that this four parameter fit gives an excellent match to the exact equation of state. Table 3.1 gives the best fit values for the equation of state parameters for the different quintessence models taken from Corasaniti and Copeland (2003), with the addition of the INV2 model. The parametrization for the dark energy equation of state is plotted in Fig. 3.1 for the various quintessence models used in this chapter.

Figure 3.2 shows the evolution of the dark energy density with expansion factor in each quintessence model. Some of these models display significant levels of dark energy at high redshifts in contrast to a ΛCDM cosmology. As the AS, CNR, 2EXP and SUGRA models have non-negligible dark energy at early times, all of these could be classed as 'early dark energy' models. As shown in Fig. 3.2 both the CNR and the 2EXP models have high levels of dark energy at high redshifts compared to ΛCDM; after an early rapid transition, the dark energy density evolves in the same way as in a ΛCDM cosmology. Other models, like the AS, INV1 and the SUGRA models, also have non-negligible amounts of dark energy at early times, and after a late-time transition, the dark energy density mimics a ΛCDM cosmology at very low redshifts. In Sect. 3.3 we will investigate if quintessence models which feature an early or late transition in their equation of state, and in their dark energy density, can be distinguished from ΛCDM by examining the growth of large scale structure. The luminosity distance and Hubble parameter in the quintessence models are compared to ΛCDM in Figs. 3.3 and 3.4, respectively. In these plots it is clear that the CNR and the 2EXP models differ from ΛCDM only at very high redshifts. The adoption of a 4 variable parametrization is essential to accurately model the expansion history over the full range of redshifts probed by the simulations. Using a 1 or 2 parameter equation of state whose application is limited to low redshift measurements restricts the analysis of the properties of dark energy and cannot make use of high redshift measurements such as the CMB. As an example, Corasaniti (2004) demonstrated that a two parameter log expansion for $w(z)$ proposed by Gerke and Efstathiou (2002), can only take into account a quintessence model which varies slowly and cannot faithfully reproduce the original $w(z)$ at high redshifts. Bassett et al. (2004) analysed how accurately various parametrizations could reproduce the dynamics of quintessence models. They found that parametrizations based on an expansion to first order in z or $\log z$ showed errors of $\sim 10\%$ at $z = 1$. A general prescription for $w(z)$ containing more parameters than a simple 1 or 2 variable equation can

Table 3.1 The equation of state of the dark energy models simulated, expressed in the parametrization of Corasaniti and Copeland (2003)

Model	w_0	w_m	a_m	Δ_m
INV1	-0.4	-0.27	0.18	0.5
INV2	-0.79	-0.67	0.29	0.4
SUGRA	-0.82	-0.18	0.1	0.7
2EXP	-1.0	0.01	0.19	0.043
AS	-0.96	-0.01	0.53	0.13
CNR	-1.0	0.1	0.15	0.016

The evolution of $w(a)$ is described by four parameters, the value of the equation of state today, w_0, and during matter domination era, w_m, the expansion factor, a_m, when the field changes its value during matter domination and the width of the transition, Δ_m. We have added the INV2 model to this list as an example of an inverse power law potential with a value of w_0 closer to -1 than in the INV1 model

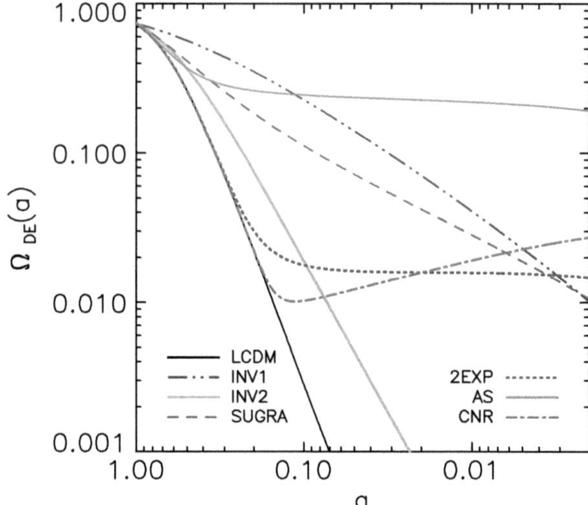

Fig. 3.2 The dark energy density, $\Omega_{DE}(a)$, as a function of expansion factor. The INV1, SUGRA, CNR, 2EXP and AS models have significant levels of dark energy at early times. From $z \sim 9$ until today the 2EXP and CNR models display the same energy density as ΛCDM. Note the x-axis scale on this plot goes to $z > 300$ on the *right hand side*

accurately describe both slowly and rapidly varying equations of state (Bassett et al. 2004). For example, the parametrization provided by Corasaniti and Copeland (2003) can accurately mimic the exact time behaviour of $w(z)$ to $<5\%$ for $z < 10^3$ using a 4 parameter equation of state and to $<9\%$ for $z < 10^5$ with a 6 parameter equation. Finally, we note that the parametrization for w proposed by Corasaniti and Copeland (2003) is similar to the four parameter equation of state in Linder and Huterer (2005) (Model 4.0) where the evolution of w is described in terms of the e-fold variable, $N = \ln a$, where a is the scale factor.

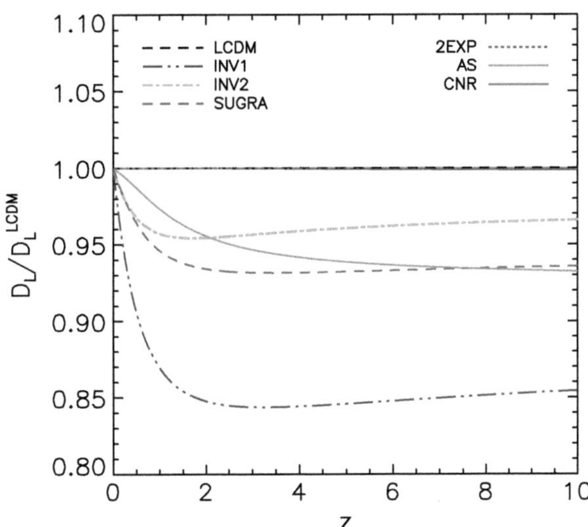

Fig. 3.3 The luminosity distance in different quintessence models compared to that in a ΛCDM cosmology. In this case we have assumed the same matter density today of $\Omega_m = 0.26$ in each of the models. The CNR and 2EXP models predict the same D_L as in ΛCDM and are overplotted

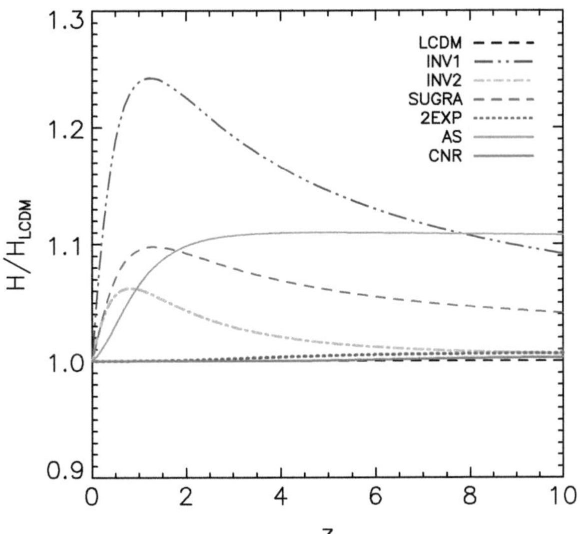

Fig. 3.4 The ratio of the Hubble parameter for quintessence cosmologies to that in ΛCDM

3.2.2 *The Expected Impact of Dark Energy on Structure Formation*

The growth of structure is sensitive to the amount of dark energy, as this changes the rate of expansion of the Universe. As a result, a quintessence model with a varying equation of state could display different large scale structure from a ΛCDM model. Varying the equation of state will result in different amounts of dark energy at different times. It has been shown that models with a larger density of dark energy at high redshift than ΛCDM have more developed large scale structure at early times, when normalised to the same σ_8 today (Grossi and Springel 2009; Francis et al. 2008).

The normalised growth factor $G = D/a$ obeys the following evolution equation in dark energy cosmologies (Linder and Jenkins 2003),

$$G'' + \left(\frac{7}{2} - \frac{3}{2}\frac{w(a)}{1 + X(a)}\right)\frac{G'}{a} + \frac{3}{2}\frac{1 - w(a)}{1 + X(a)}\frac{G}{a^2} = 0, \tag{3.3}$$

where

$$X(a) = \frac{\Omega_{\mathrm{m}}}{1 - \Omega_{\mathrm{m}}}\, e^{-3\int_a^1 \mathrm{d}\ln a'\, w(a')}, \tag{3.4}$$

$w(a)$ is the dynamical dark energy equation of state and a prime denotes a derivative with respect to the scale factor. The linear growth factor for each quintessence model is plotted in Fig. 3.5. In Sect. 3.3.1, we present the simulation results for each quintessence model where the initial conditions were generated using a ΛCDM linear theory power spectrum and the background cosmological parameters are the best fit values assuming a ΛCDM cosmology (Stage I). The difference between the simulations is the result of having a different linear growth rate for the dark matter perturbations.

The presence of small but appreciable amounts of dark energy at early times also modifies the growth rate of fluctuations from that expected in a matter dominated universe and hence changes the shape of the linear theory $P(k)$ from the ΛCDM prediction. The quintessence scalar field can contribute at most a small fraction of the total energy density at early redshifts. Constraints on this amount come from big bang nucleosynthesis as well as from CMB measurements. Bean et al. (2001) found a limit of $\Omega_{\mathrm{DE}} < 0.045$ at $a \sim 10^{-6}$ using the observed abundances of primordial nuclides and a constraint of $\Omega_{\mathrm{DE}} < 0.39$ during the radiation domination era, $a \sim 10^{-4}$, from CMB anisotropies. Caldwell et al. (2003) discuss the parameter degeneracies which allow for different amounts of dark energy at early times leaving the position of the CMB peaks unchanged (see Sect. 3.3.3). Using the WMAP first year data, Corasaniti et al. (2004) found a limit of $\Omega_{\mathrm{DE}} < 0.2$ at $z \sim 10$. Some recent parametrization dependent constraints on early dark energy models found the dark energy density parameter to be $\Omega_{\mathrm{DE}} < 0.02$ at the last scattering surface (Xia and Viel 2009). Note that all of the models we consider are consistent with this constraint, except for the AS model (see Fig. 3.2).

If the dark energy is not a cosmological constant, then there will be dark energy perturbations present, δ_φ whose evolution will affect the dark matter power spectrum and alter the evolution equation in Eq. 3.3 (Ferreira and Joyce 1998; Weller and Lewis 2003). As most of the quintessence models we will consider display a non-negligible contribution to the overall density from dark energy at early times, the matter power spectrum is affected in two ways (Ferreira and Joyce 1998; Caldwell et al. 2003; Doran et al. 2007). In the matter dominated era, the growing mode solution for

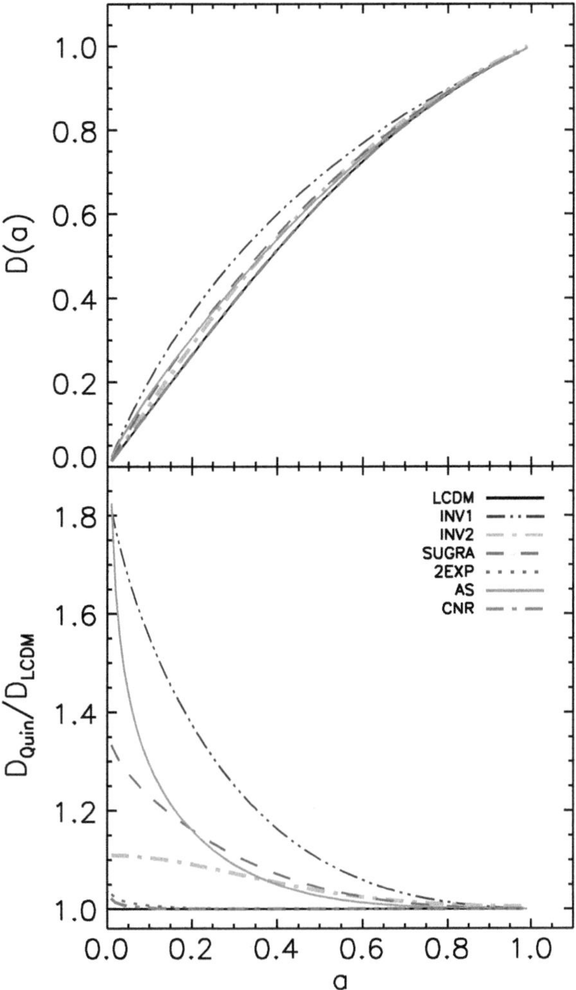

Fig. 3.5 The growth factor as a function of expansion factor. The *upper panel* shows the evolution of the linear growth factor in each quintessence model. In the *lower panel* the ratio of the growth factor in the quintessence models compared to ΛCDM is plotted. The growth factor in each case has been normalised to unity today

dark matter density perturbations is proportional to the expansion factor, $\delta_{\mathrm{m}} \propto a$, in a universe without a scalar field component. In a dark energy model which has appreciable amounts of dark energy at early times, the dark matter growing mode solution on subhorizon scales is modified to become

$$\delta_{\mathrm{m}} \propto a^{[\sqrt{25-24\Omega_{\mathrm{DE}}}-1]/4}. \tag{3.5}$$

The growth of modes on scales $k > k_{\mathrm{eq}}$, where k_{eq} is the wavenumber corresponding to the horizon scale at matter radiation equality, is therefore suppressed relative to the growth expected in a ΛCDM universe. For fluctuations with wavenumbers $k < k_{\mathrm{eq}}$ during the matter dominated epoch, the suppression takes place after the mode enters the horizon and the growing mode is reduced relative to a model with $\Omega_{\mathrm{DE}} \simeq 0$. These

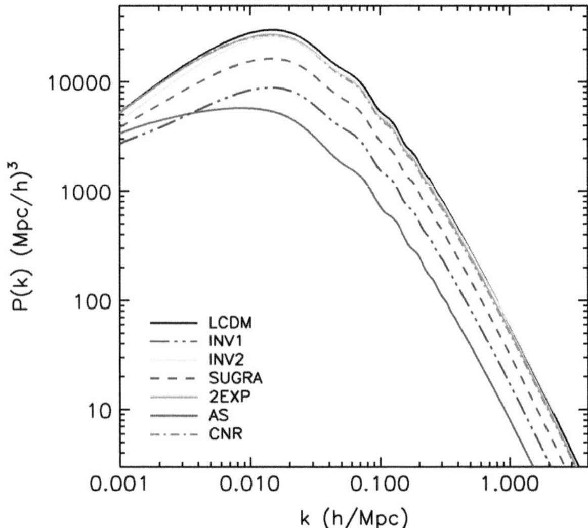

Fig. 3.6 Linear theory power spectra at $z = 0$ for dynamical dark energy quintessence models and ΛCDM. In this plot, the spectra are normalised to CMB fluctuations (on smaller wavenumbers than are included in the plot). The presence of a non-negligible dark energy density fraction at early times causes a scale independent suppression of growth for scales $k > k_{eq}$ where k_{eq} is the wavenumber corresponding to the horizon scale at matter radiation equality and a scale dependent suppression at $k < k_{eq}$. Models with high Ω_{DE} at the last scattering surface have a lower σ_8 today compared to ΛCDM if normalised to CMB fluctuations

two effects are illustrated for a scaling quintessence model in Ferreira and Joyce (1998), whose Fig. 7 shows the evolution of δ_m for two wavenumbers, one that enters the horizon around a_{eq} ($k = 0.1\,\mathrm{Mpc}^{-1}$) and one that comes in during the radiation era ($k = 1\,\mathrm{Mpc}^{-1}$), in a universe with $\Omega_{DE} = 0.1$ during the matter dominated era. There is a clear suppression of growth after horizon crossing compared to a universe with no scalar field. The overall result is a scale independent suppression for subhorizon modes, a scale dependent red tilt ($n_s < 1$) for superhorizon modes and an overall broading of the turnover in the power spectrum. This change in the shape of the turnover in the matter power spectrum can be clearly seen in Fig. 3.6 for the AS model. This damping of the growth after horizon crossing will result in a smaller σ_8 value for the quintessence models compared to ΛCDM if normalised to CMB fluctuations (see also Kunz et al. 2004).

We have used the publicly available Parametrized Post-Friedmann (PPF) module for CAMB (Fang et al. 2008) to generate the linear theory power spectrum. This module supports a time dependent dark energy equation of state by implementing a PPF prescription for the dark energy perturbations with a constant sound speed $c_s^2 = 1$. Figure 3.6 shows the dark matter power spectra at $z = 0$ generated by CAMB for each quintessence model and ΛCDM with the same cosmological parameters, an initial scalar amplitude of $A_s = 2.14 \times 10^{-9}$ and a spectral index $n_s = 0.96$ (Sánchez et al. 2009). As can be seen in this plot, models with higher fractional energy densities at early times have a lower σ_8 today and a broader turnover in $P(k)$. In Sect. 3.3.2 a consistent linear theory power spectrum was used for each quintessence model to generate the initial conditions for the simulations (Stage II).

Finally, quintessence dark energy models will not necessarily agree with observational data when adopting the cosmological parameters derived assuming a ΛCDM cosmology. We consider how the different quintessence models affect various distance scales. We find the best fit cosmological parameters for each quintessence model using the observational constraints on distances such as the measurements of the angular diameter distance and sound horizon at the last scattering surface from the cosmic microwave background. The method and data sets used are given in Appendix A.1 and the corresponding simulation results which use a consistent linear theory power spectrum for each model together with the best fit cosmological parameters are presented in Sect. 3.3.3 (Stage III).

3.2.3 Simulation Details

For each of the quintessence models the parametrization for the dark energy equation of state given in Eq. 3.2 was used. In the first stage we fix the cosmological parameters for all of the quintessence models to those of ΛCDM. As a result, some of the scalar field models do not match observational constraints on the sound horizon at last scattering or the angular diameter distance. We shall discuss this further in Sect. 3.3.3 using the results given in Appendix A.1. In the first stage of our calculations, presented in Sect. 3.3.1, the linear theory power spectrum used to set up the initial conditions in the quintessence models was the same as ΛCDM. For the purpose of computing the shape of $P(k)$ in Stage I, we have assumed that the ratio of dark energy density to the critical density at the last scattering surface ($z_{\rm lss} \sim 1000$) is negligible and have ignored any clustering of the scalar field dark energy. To generate the initial conditions for the simulations with dynamical dark energy, the growth factor, which appears in the Zel'dovich approximation, needs to be computed numerically using the growth equation in Eq. 3.3. In Sect. 3.3.2, the linear theory $P(k)$ is generated for each quintessence model using a modified version of CAMB which incorporates the influence of dark energy on dark matter clustering at early times. In each model the power spectra at redshift zero have been normalised to have $\sigma_8 = 0.8$. Using the linear growth factor for each dark energy model, the linear theory $P(k)$ was then evolved backwards to the starting redshift of $z = 200$ in order to generate the initial conditions for `L-Gadget-2`. The power spectrum was computed by assigning the particles to a mesh using the cloud in cell (CIC) assignment scheme (Hockney and Eastwood 1981) and performing a fast Fourier transform of the density field. To compensate for the mass assignment scheme we perform an approximate de-convolution following Baumgart and Fry (1991). Snapshot outputs of the dark matter distribution as well as the group catalogues were made at redshifts 5, 3, 2.5, 2, 1.5, 1, 0.75, 0.5, 0.25 and 0. We investigate gravitational collapse in the six quintessence models listed in Table 3.1 by comparing the evolution of the power spectrum at various redshifts.

3.3 Results

In the following sections we present the power spectrum predictions from the three stages of simulations carried out as described in Sect. 3.2. The bottom line results are presented in Sect. 3.3.3, in which we compare power spectra in ΛCDM with a subset of dark energy models which also pass the currently available observational constraints. Sections 3.3.1 and 3.3.2 show intermediate steps away from ΛCDM towards the consistent dark energy models presented in Sect. 3.3.3, to allow us to understand the impact on $P(k)$. In Sect. 3.3.1 the Friedmann equation was modified with the quintessence model's equation of state as a function of redshift and a ΛCDM linear theory power spectrum was used to generate the initial conditions for all the simulations (Stage I). In Sect. 3.3.2 we use a consistent linear theory power spectrum for each quintessence model (Stage II). In Sect. 3.3.3 we constrain a set of cosmological parameters, using CMB, BAO and SN data, for each dark energy model. The final stage of simulations use a consistent linear theory power spectrum for each model together with the best fit cosmological parameters (Stage III).

3.3.1 Stage I: Changing the Expansion Rate of the Universe

In this first stage of simulations, the same ΛCDM initial power spectrum and cosmological parameters were used for all models. In Fig. 3.7 we plot the power spectrum at redshifts $z = 0, 1, 5$ in ΛCDM (orange lines) and in the AS model (green lines), together with the linear theory power spectra for ΛCDM (black lines). The AS model has a linear growth rate that differs from ΛCDM by $\sim$20 % at $z = 5$. We also plot the Smith et al. (2003) 'Halofit' empirical fitting function for ΛCDM and the AS model. The Halofit function has been incorporated into the CAMB package and this code was used to generate the output at various redshifts seen in Fig. 3.7. As this plot shows, the Smith et al. (2003) expression accurately describes the evolution of the power spectrum at redshift 0 in both models and at earlier times. As the normalisation and linear spectral shape is the same in these two models, Halofit accurately reproduces the nonlinear power in each model at various redshifts once the appropriate linear growth factor for the dark energy model at that redshift is used. The Smith et al. expression agrees with the simulation output at $z = 0$ to within 4 % for $k < 1\,\mathrm{h\,Mpc^{-1}}$ for both the quintessence model and ΛCDM. At higher redshifts, the difference between the simulation output and the Halofit prediction for all the models is just under 10 % on scales $k < 0.3\,\mathrm{h\,Mpc^{-1}}$ at $z = 5$.

To highlight the differences in the power between the different models, we plot in Fig. 3.8 the measured power divided by the power at $z = 5$, after scaling to take into account the difference in the linear theory growth factors for the output redshift and $z = 5$, for ΛCDM. This removes the sampling variance from the plotted ratio (Baugh and Efstathiou 1994). A ratio of unity in Fig. 3.8 would indicate linear growth at the same rate as expected in ΛCDM.

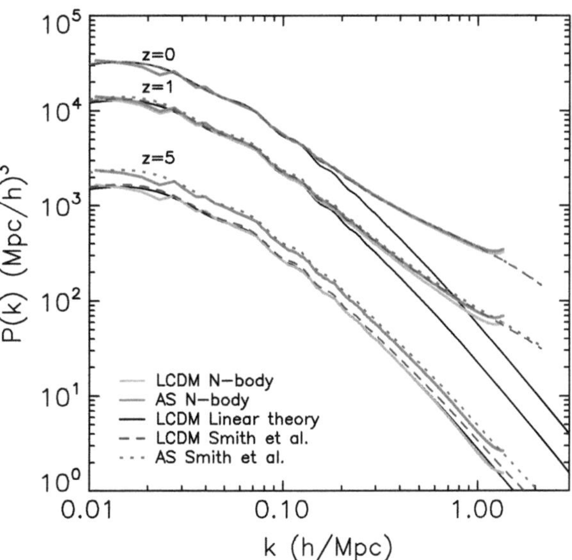

Fig. 3.7 Power spectra in a ΛCDM cosmology (*orange lines*) and AS quintessence model (*green lines*) at redshift 0, 1 and 5. The *dashed lines* correspond to the (Smith et al. 2003) analytical expression for the nonlinear $P(k)$ in ΛCDM; *dotted lines* show the equivalent for the AS model. The *solid black line* is the linear theory for ΛCDM at the corresponding redshift outputs. The Smith et al. (2003) expression for the AS model has been scaled with the appropriate growth factor for this model at each redshift (colour figure online)

Figure 3.8 shows four epochs in the evolution of the power spectrum for all of the quintessence models and ΛCDM. The black line in the plot shows the $P(k)$ ratio for ΛCDM (note the yellow curve for the CNR model is overplotted). Non-linear growth can be seen as an increase in the power ratio on small scales, $k > 0.3\,\mathrm{h\,Mpc}^{-1}$ at $z = 3$ and $k > 0.1\,\mathrm{h\,Mpc}^{-1}$ at $z = 0$. Four of the quintessence models (INV1, INV2, SUGRA and AS) differ significantly from ΛCDM for $z > 0$. These models show advanced structure formation i.e. more power than ΛCDM, and a large increase in the amount of nonlinear growth. All models are normalised to have $\sigma_8 = 0.8$ today and as a result all the power spectra are very similar at redshift zero in Fig. 3.8. There are actually small differences between the quintessence models at $z = 0$ as seen on the expanded scale in Fig. 3.9. This increase in nonlinear power at small scales in the quintessence models is due to the different growth histories.

The power spectra predicted in the 2EXP and CNR models show minor departures from that in the ΛCDM cosmology. This is expected as Figs. 3.1 and 3.2 show the equations of state and the dark energy densities in these two models are the same as ΛCDM at low redshifts and all three simulations began from identical initial conditions. It could be possible to distinguish these two models from the concordance cosmology at higher redshifts if we do not ignore the dark energy perturbations or changes in the growth factor which alter the form of the linear theory power spectrum. We shall discuss this more in the next stage of our simulations in Sect. 3.3.2.

Finally, we investigate if the enhanced growth in the power spectrum seen in Fig. 3.8 in the quintessence models is due solely to the different linear growth rates at a given redshift in the models. In order to test this idea, the power spectrum in a quintessence model and ΛCDM are compared not at the same redshift but at the

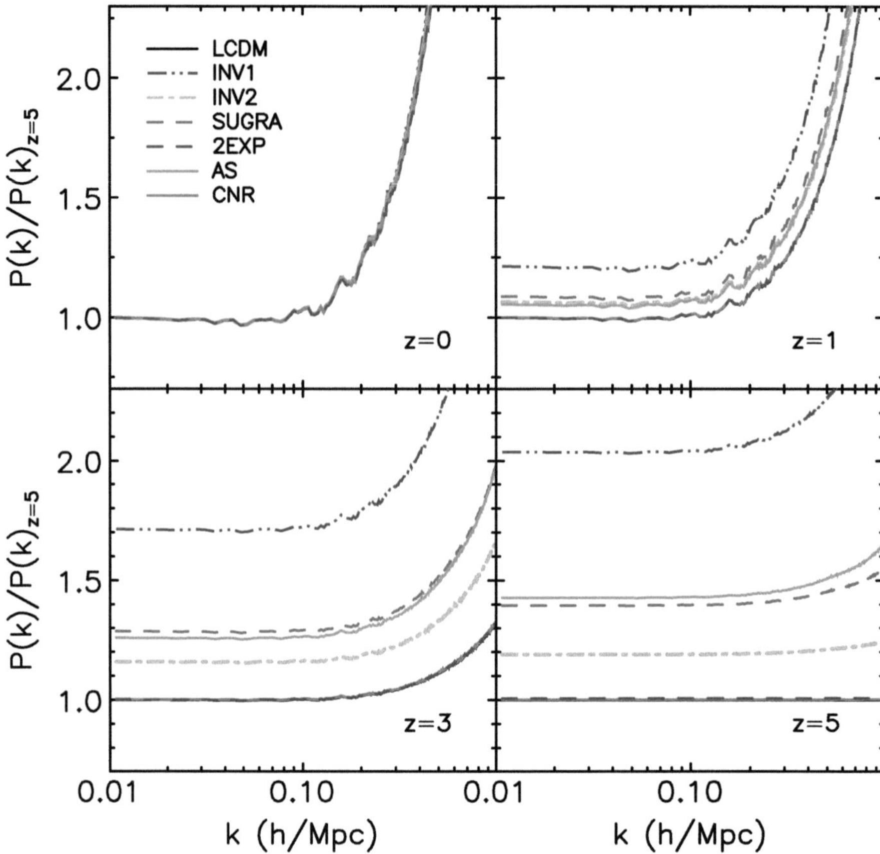

Fig. 3.8 The nonlinear growth of the power spectra in the various quintessence models as indicated by the key in the *top left panel*. Each panel shows a different redshift. The power spectra in each case have been divided by the ΛCDM power spectrum at redshift 5 scaled to take out the difference between the ΛCDM growth factor at $z = 5$ and the redshift plotted in the panel. This removes the sampling variance due to the finite box size and highlights the enhanced nonlinear growth found in quintessence cosmologies compared to ΛCDM. A deviation of the power ratio from unity therefore indicates a difference in $P(k)$ from the linear perturbation theory of ΛCDM (colour figure online)

same linear growth factor.[1] As the growth rates in some of the quintessence models are very different from that in the standard ΛCDM cosmology, the power spectra required from the simulation will be at different output redshift in this comparison. For example, the normalised linear growth factor is $D = 0.5$ at a redshift of $z = 1.58$ in a ΛCDM model and has the same value at $z = 1.82$ in the SUGRA model, at $z = 1.75$ in the AS model and at $z = 2.25$ in the INV1 quintessence model. In Fig. 3.10 we show the power spectrum of simulation outputs from the INV1, AS, SUGRA and CNR models divided by the power spectrum output in ΛCDM at the

---

[1] We thank S.D.M. White for this suggestion.

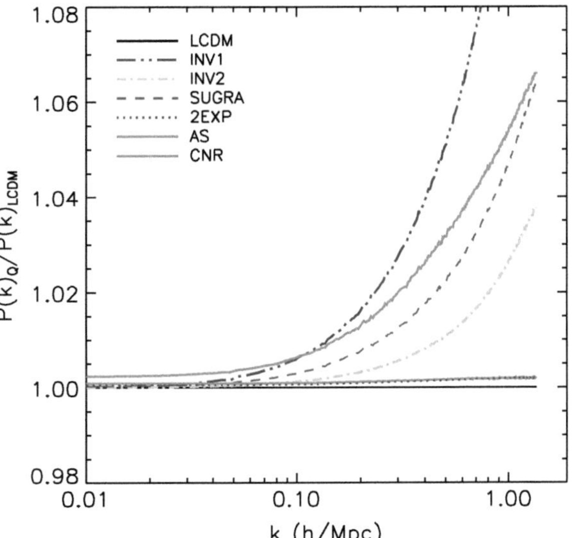

Fig. 3.9 Ratio of power spectra output from the simulations in the six quintessence models compared to the nonlinear ΛCDM $P(k)$ at redshift 0. Note the expanded scale on the y-axis. As expected, the 2EXP and CNR models show no difference from ΛCDM while the difference in the INV1, INV2, SUGRA and AS models is under 10 % for wavenumbers $k < 1\,\mathrm{h\,Mpc^{-1}}$

same linear growth rate. We ran the simulations taking three additional redshift outputs where the linear growth rate had values of $D = 1$, $D = 0.5$ and $D = 0.3$. It is clear from Fig. 3.10 that scaling the power spectrum in this way can explain the enhanced linear and most of the excess nonlinear growth seen in Fig. 3.8 for scales $k < 0.1\,\mathrm{h\,Mpc^{-1}}$. For example, in the INV1 model the enhanced nonlinear growth, on scales $k \sim 0.3\,\mathrm{h\,Mpc^{-1}}$ at fixed $D = 0.3$, differs from ΛCDM by at most 5 % in Fig. 3.10 as opposed to at most 30 % at $z = 5$ in Fig. 3.8. At earlier redshifts when the linear growth rate is $D = 0.3$, the nonlinear growth in the quintessence models agrees with ΛCDM on smaller wavenumbers $k < 0.3\,\mathrm{h\,Mpc^{-1}}$. As in Fig. 3.8, the CNR model shows no difference from ΛCDM when plotted in this way.

Note in Fig. 3.10 the INV1 model has less nonlinear growth at $D = 0.3$ and $D = 0.5$ compared to the AS model. The AS and SUGRA models have a growth rate of $D = 0.5$ at lower redshifts compared to the INV1 model and so are at a later stage in their growth history. The INV1 model has a growth rate of $D = 0.5$ at $z = 2.25$ whereas for the AS model this occurs at $z = 1.75$ and at 1.82 for the SUGRA model. The reason for the success of this simple model—matching the growth factor to predict the clustering—can be traced to the universality of the mass function, which we discuss in Sect. 3.3.4. In this Stage I calculation, the models have the same mass function when plotted at the epoch corresponding to a common growth factor. This means that the two-halo contribution to the clustering is therefore the same. Can this simple halo picture of the clustering also explain the clustering on small scales (high k)? Although the abundance of haloes in the models is the same at the epochs corresponding to a given value of the growth factor, the concentrations of the haloes will not be the same. In cosmologies where the haloes formed at a higher redshift (i.e. roughly the redshift corresponding to a particular value of D),

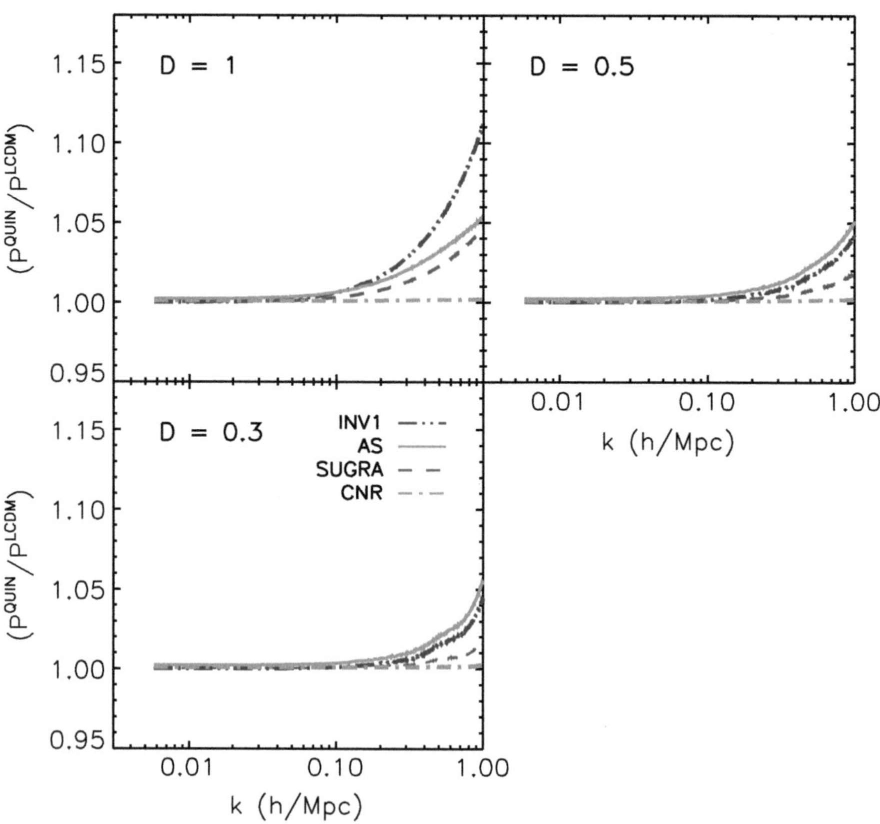

Fig. 3.10 The ratio of the quintessence model power spectra to the ΛCDM power spectrum output from the simulations at three values of the linear growth factor $D = 1$, $D = 0.5$ and $D = 0.3$. Each panel shows the results of this exercise for the AS, CNR, 2EXP and SUGRA quintessence models. The growth factors correspond to $z = 3.4$ ($D = 0.3$), $z = 1.6$ ($D = 0.5$) and $z = 0$ ($D = 1$) for ΛCDM. For each model, the choice of growth factor corresponds to slightly different redshifts, with the biggest difference being for the INV1 model. A ratio of unity would indicate that the growth factor is the only ingredient needed to predict the power spectrum in the different quintessence models. Note the expanded scale on the y-axis

one would expect these haloes to have higher concentrations than their counterparts in the other models (Eke et al. 2001). A higher concentration would be expected to yield stronger nonlinear clustering and hence more power at high k in Fig. 3.10. Unfortunately our simulations do not have the resolution to probe the required range of wavenumbers to uncover this behaviour. The ratios plotted in Fig. 3.10 stop at wavenumbers approximately equivalent to the collapsed radius of a massive halo.

Hence, it seems that scaling the power spectrum using the linear growth rate can be used to predict the linear growth in the quintessence dark energy simulations and can reproduce some of the nonlinear growth at early redshifts. In Fig. 3.10 there are still some differences in the small scale growth in quintessence models compared to

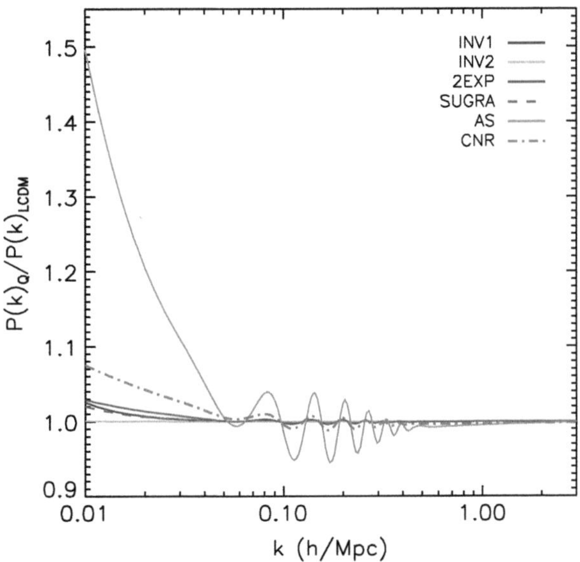

Fig. 3.11 Ratio of linear theory power spectra for quintessence models shown in Fig. 3.6 to that in ΛCDM. In this plot each $P(k)$ has been normalised so that $\sigma_8 = 0.8$ today; this is the normalisation used in our simulations

ΛCDM which cannot be explained by the different linear growth rates. We find that nonlinear evolution is not just a function of the current value of the linear growth rate but also depends on its history through the evolution of the coupling between long and short-wavelength modes.

3.3.2 Stage II: Use of a Self-Consistent Linear Theory P(k)

We have run the simulations presented in the previous section again but this time using the appropriate linear theory $P(k)$ for each model (shown in Fig. 3.6) normalised to $\sigma_8 = 0.8$ today (Stage II). After normalising the power spectra in this way, the difference between the quintessence models $P(k)$ and ΛCDM can be seen in Fig. 3.11. The INV2 model was not included in this set of simulations as there is a negligible difference in the linear theory power spectrum from ΛCDM. Note Francis et al. (2008) also generate the linear theory power spectrum for 'early dark energy' models and normalise all $P(k)$ to have the same σ_8 today. Francis et al. (2008) make an equivalent plot to Fig. 3.11 but find a decrease in this ratio with decreasing scale ($k > 0.2\,\mathrm{h\,Mpc}^{-1}$), using the parametrization for early dark energy proposed by Doran and Robbers (2006), in contrast to the ratio of unity we find on small scales in Fig. 3.11. This difference is due to the different parametrizations used for the dark energy equation of state, as a ratio of unity is obtained on small scales for the same 'early dark energy' model using the parametrization suggested by Wetterich (2004) (M. Francis, private communication).

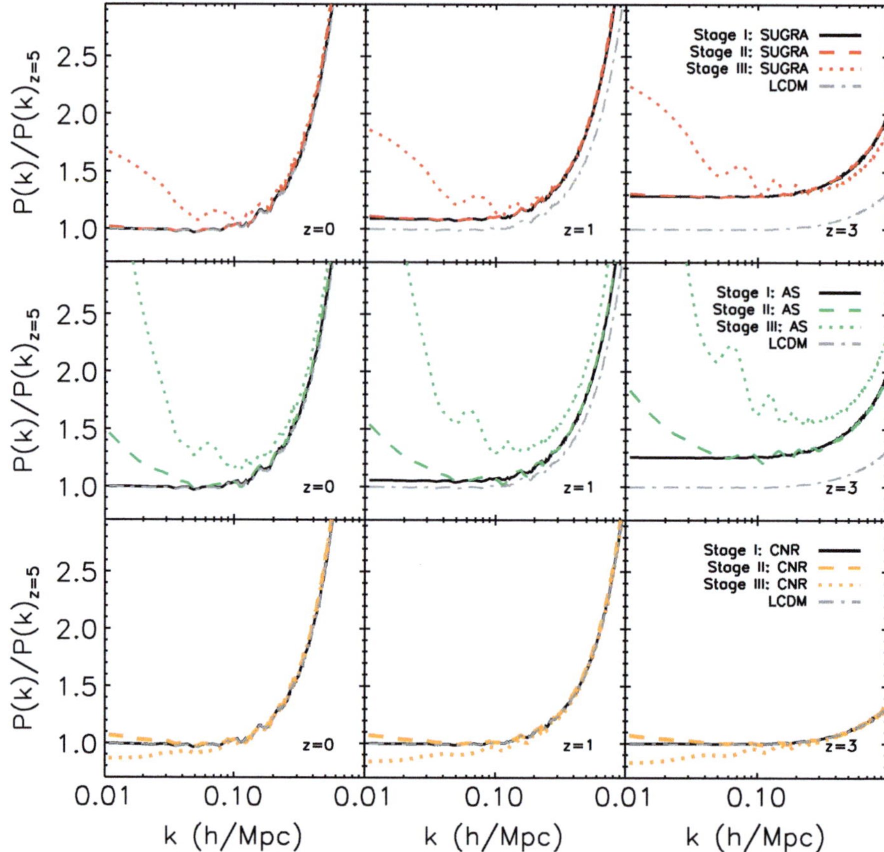

Fig. 3.12 Ratios of power spectra for the SUGRA (*first row*), AS (*second row*) and CNR (*third row*) quintessence model compared to ΛCDM from the three stages of simulations in this chapter. The plot shows the growth in the quintessence models using ΛCDM linear theory $P(k)$ in the initial conditions in black (Stage I) and using a self consistent linear theory $P(k)$ for each quintessence model (*dashed colored line*) (Stage II). The *dotted lines* show the $P(k)$ ratio from the simulation for the quintessence models using the best fit parameters in Table A.1 (Stage III). The power spectra in each case have been divided by the ΛCDM power spectrum at redshift 5 with appropriate scaling of ΛCDM growth factors. The linear theory power spectra in each case has been normalised to $\sigma_8 = 0.8$

In the first row of Fig. 3.12 we plot the power spectrum for the Stage II SUGRA model at $z = 0$, 1, and 3 divided by the simulation output in ΛCDM at $z = 5$ as in Fig. 3.8 (red dashed lines). The result from Fig. 3.8, Stage I SUGRA, is also plotted here to highlight how changing the spectral shape affects the nonlinear growth in the simulations. On large scales the growth is not modified by the altered spectral shape. The growth of perturbations on small scales in the simulation is affected by the modified linear theory used in the initial conditions. Normalising the power spectra to $\sigma_8 = 0.8$ results in more power on large scales in the quintessence models compared to ΛCDM, as can be seen in Fig. 3.11. This enhanced large scale power couples to the power on smaller scales and results in a small increase in the nonlinear power spectrum for $k > 0.1\,\mathrm{h\,Mpc}^{-1}$ in the Stage II SUGRA simulation compared to the one using ΛCDM linear theory $P(k)$ in Stage I.

In the second row of Fig. 3.12 we plot the power spectrum for the Stage II AS model as green dashed lines at $z = 0$, 1, and 3 divided by the simulation output in ΛCDM at $z = 5$ as in Fig. 3.8. The growth of dark matter perturbations is greatly suppressed in the AS model due to the large fractional dark energy density at high redshifts. After fixing $\sigma_8 = 0.8$, there is more power on large scales in the AS model compared to ΛCDM. As in the first row of Fig. 3.12 there is a small increase in nonlinear power for the AS model in Stage II. Although the excess large scale power is significantly larger than in the SUGRA model case, it does not result in more nonlinear power on small scales through mode coupling, as can be seen in the panels in the second row in Fig. 3.12. The linear theory power spectrum for these quintessence models has a scale dependent red tilt on large scales which shifts the position of the BAO peaks which is the origin of the oscillation apparent in the second row of Fig. 3.12 at $z = 3$. The difference in BAO peak positions is very prominent when we plot the ratio of the power spectrum in the AS model to the ΛCDM power spectrum and can be clearly seen in Fig. 3.12.

3.3.3 Stage III: Consistency with Observational Data

In this section we present the power spectra results in ΛCDM and a subset of the dark energy models, measured from simulations which use a consistent linear theory power spectrum for each model together with the best fit cosmological parameters. We have simulated the SUGRA, AS and CNR models using the best fit cosmological parameters from Table A.1 and the linear theory power spectrum specific to each model as discussed in Sect. 3.2.2. We chose to simulate these three models following the analysis and results of Sects. 3.3.1, 3.3.2 and Appendix A.1. Any of the dark energy models listed in Sect. 3.2 which showed similar results in Sect. 3.3.2 to ΛCDM and similar cosmological parameters in Appendix A.1 have not been simulated again. Table A.1 in Appendix A.1 shows the best fit values for $\Omega_m h^2$, $\Omega_b h^2$ and H_0 for each quintessence model, found by minimising $\chi^2_{\mathrm{total}} = \chi^2_{\mathrm{WMAP+SN+BAO}}$. The SUGRA, AS and CNR models had the biggest improvement in the agreement with observational constraints, on allowing $\Omega_m h^2$, $\Omega_b h^2$ and H_0 to vary. The results for the SUGRA, AS and the CNR model are shown as dotted coloured lines in Fig. 3.12 and are referred to as Stage III in the legend to distinguish them from the results of Sects. 3.3.1 and 3.3.2 which are also plotted. In each row we show the simulation outputs at $z = 0$, 1 and 3. The simulation results for each quintessence model uses the models linear theory and the best fit parameters from Table A.1. Using the best fit parameters for each model together with the correct linear theory changes the growth of structure in the simulation.

In Fig. 3.12 the measured power spectrum for each model is divided by the power for ΛCDM at $z = 5$ which has been scaled using the difference in the linear growth factor between $z = 5$ and the redshift shown. Plotting the ratio in this way highlights the differences in growth between the quintessence models and ΛCDM as well as removing sampling variance.

The measured power for the SUGRA model is plotted in the first row in Fig. 3.12. The power spectra have all been normalised to $\sigma_8 = 0.8$ resulting in a large increase in the large scale power ($k < 0.1\,h\,\mathrm{Mpc}^{-1}$) seen in Fig. 3.12 compared to ΛCDM. There is a large increase in the linear and nonlinear growth in this model at $z > 0$ (dotted red line) compared to ΛCDM (dot-dashed grey line). The second row in Fig. 3.12 shows there is a significant enhancement in the growth in the AS power spectrum measured compared to ΛCDM for $z < 3$. The power measured from the simulations of the CNR model are plotted in the third row of Fig. 3.12. We find there is a small reduction in the amount of linear and nonlinear growth in this model compared to ΛCDM.

In Fig. 3.12 we also plot the simulation results for these three models from Sect. 3.3.1 (Stage I), where ΛCDM linear theory was used in the initial conditions, (black lines). The dashed coloured lines show the simulation results from Sect. 3.3.2 (Stage II), where the quintessence model linear theory was used. The SUGRA power spectrum measured in Stage III has less nonlinear growth at high redshifts compared to the SUGRA $P(k)$ from Stage I or II due to changes in the spectral shape. The measured power for the AS model using the best fit parameters (Stage III) shows enhanced growth on all scales compared to the power for the AS model in Stage I (using ΛCDM parameters and linear theory $P(k)$) or Stage II (using ΛCDM parameters).

These results show the importance of each of the three stages in building up a complete picture of a quintessence dark energy model. Models whose equation of state is very different from ΛCDM at low redshifts, for example the SUGRA and the AS model, show enhanced nonlinear growth today compared to ΛCDM. Models whose equation of state is very different to ΛCDM only at early times, for example the CNR model, will show no difference in the nonlinear growth of structure if we use the ΛCDM spectral shape (Stage I). In Stage II and III the shape of the power spectrum in the CNR model has changed and is very different to ΛCDM on large scales as can be seen in Fig. 3.12. Using the best fit cosmological parameters for this model we find a very small reduction ($<2\,\%$) in the nonlinear growth at $z = 0$ compared to ΛCDM.

3.3.4 Mass Function of Dark Matter Haloes

In this section we present the mass function of dark matter haloes in the quintessence models using the three stages of simulations discussed in Sects. 3.3.1–3.3.3.

Press and Schechter (1974) (hereafter P-S) proposed an analytical expression for the abundance of collapsed objects with mass M in the range M to $M + dM$ at redshift z, based on the spherical collapse model in which a perturbation can be associated with a virilised object at $z = z'$, if its density contrast, extrapolated to $z = z'$ using linear theory, exceeds some threshold value, δ_c, the critical linear density contrast. It has been shown that the P-S approach fails to reproduce the abundance of haloes found in simulations, overpredicting the number of haloes below the characteristic mass M_* and underpredicting the abundance in the high mass tail (Efstathiou and Rees 1988; White et al. 1993; Lacey and Cole 1994; Eke et al. 1996; Governato et al. 1999).

It is thought that the main cause of this discrepancy is the spherical collapse approximation, as the perturbations in the density field are inherently triaxial. After turnaround, each axis may evolve separately until the final axis collapses and the object virilises. Sheth et al. (2001) and Sheth and Tormen (2002) (hereafter S-T) modified the P-S formalism, replacing the spherical collapse model with ellipsoidal collapse, in which the surrounding shear field as well as the initial overdensity determines the collapse time of an object. Sheth et al. (2001) found a universal mass function for any CDM model. Jenkins et al. (2001) found a universal empirical fit to the form of the mass function measured from a suite of cosmological simulations. The Jenkins et al. mass function can accurately predict halo abundances over a range of cosmologies and redshifts (see also Warren et al. 2006; Reed et al. 2007; Crocce et al. 2010).

We use a friends-of-friends (FOF) halo finder, with a constant linking length of $b = 0.2$, to identify haloes in all cosmologies. In Fig. 3.13 we plot groups containing 20 particles or more to ensure that the systematic uncertainties in the mass function are at or below the 10 % level; tests show that 90 % or more of such haloes are gravitationally bound (Springel et al. 2005). The first row in Fig. 3.13 shows the mass function for SUGRA and ΛCDM at $z = 0$, 1 and 2. The filled red squares represent the mass function from Stage III of the simulations where a consistent linear theory and cosmological parameters were used for the SUGRA model. The mass function for ΛCDM (open black circles) and the SUGRA model are plotted together with the Jenkins et al. mass function shown in black (red) for ΛCDM (SUGRA). The S-T mass function is shown in the top left panel in the first row of this figure (blue dashed line) for comparison. The abundances in both ΛCDM and SUGRA agree with each other at redshift 0 and with the Jenkins et al. and S-T models, although the fitting formulae seem to slightly under-predict the number of haloes at the high mass end ($M > 10^{15} h^{-1} M_\odot$). In the first row of Fig. 3.13, the number of haloes in the two models start to differ at $z = 1$, and at $z = 2$ there is a large difference in the mass functions. The linear growth factor for the SUGRA model together with the best fit cosmological parameters from Table A.1 have been used to obtain the Jenkins et al. fit at the earlier redshifts. The Jenkins et al. fit describes the data slightly better at the high mass end at higher redshifts than the S-T prescription. This is as expected as the Jenkins et al. fit was explicitly tested at the high mass end of the mass function. Each model shows only small (<20 %) differences between the measured value and the Jenkins et al. fitting formula for $M < 10^{15} h^{-1} M_\odot$ at $z = 0$. Underneath each panel in the first row in Fig. 3.13, we plot the ratio between the measured mass function

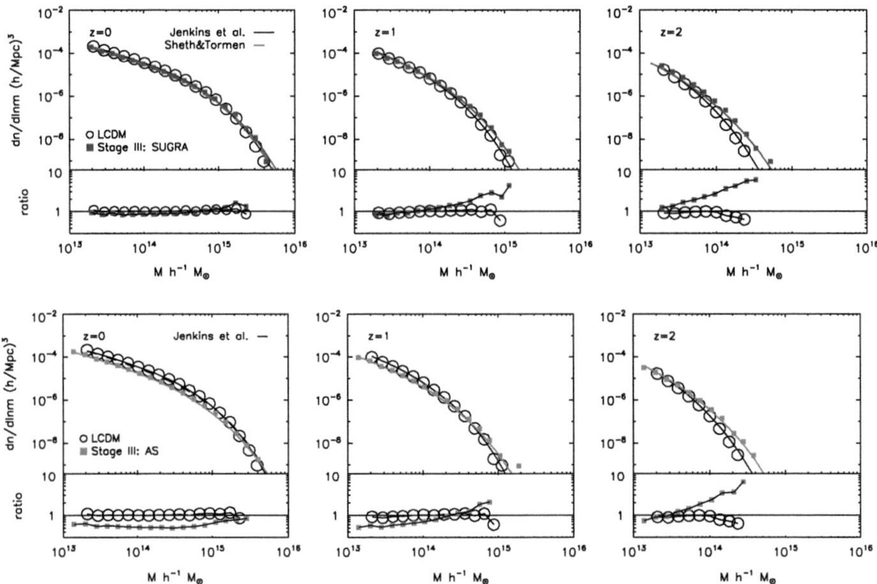

Fig. 3.13 Dark matter halo mass functions for the SUGRA (*first row*) and AS (*second row*) quintessence models compared with that in ΛCDM from the Stage III simulations at $z = 0$, 1 and 2. The mass function in ΛCDM is shown as *open black circles* throughout this plot. In the first row the *red filled squares* show the mass function from the simulation for the SUGRA model using the best fit parameters in Table A.3 (Stage III). Underneath each panel in the *first row* we plot the log of the ratio between the measured mass function for ΛCDM (*open black circles*) and Stage III SUGRA (*red squares*) and the Jenkins mass function for ΛCDM. In the *second row* the *green filled squares* show the mass function from the simulation for the AS model using the best fit parameters in Table A.3 (Stage III). For the AS Stage III simulation, $\Omega_m h^2 = 0.086$, giving rise to a change in the spectral shape of the linear theory power spectrum. As a result, there are fewer low mass halos and a similar number of high mass haloes at $z = 0$ compared to ΛCDM ($\Omega_m h^2 = 0.1334$). The difference between the Jenkins et al. mass function for ΛCDM and the measured mass function for ΛCDM (*open black circles*) and Stage III AS (*green squares*) is plotted underneath each panel in the *second row*. The *black horizontal line* indicates a ratio of unity in the ratio plots. In the *first* and *second rows* the *solid black* (*red/green*) *lines* are the predicted abundances in the ΛCDM (SUGRA/AS) model using the Jenkins et al. fitting function at various redshifts. In the *top left panel*, for reference, we have also plotted the Sheth and Tormen mass function (*blue dashed line*) for ΛCDM (colour figure online)

for ΛCDM and the SUGRA model in Stage III, and the Jenkins at al. mass function for ΛCDM.

The second row of Fig. 3.13 repeats this comparison for the AS model. In this row the mass function for ΛCDM (open black circles) and the AS model from Stage III (green squares) of the simulations at $z = 0$, 1 and 2 are plotted. The Jenkins et al. mass function for ΛCDM (black line) and the AS model for Stage III (green line) are also plotted. The AS model has a greater abundance of halos than ΛCDM at $z = 2$. For the Stage III simulation, the AS model has $\Omega_m h^2 = 0.086$ giving rise to a change

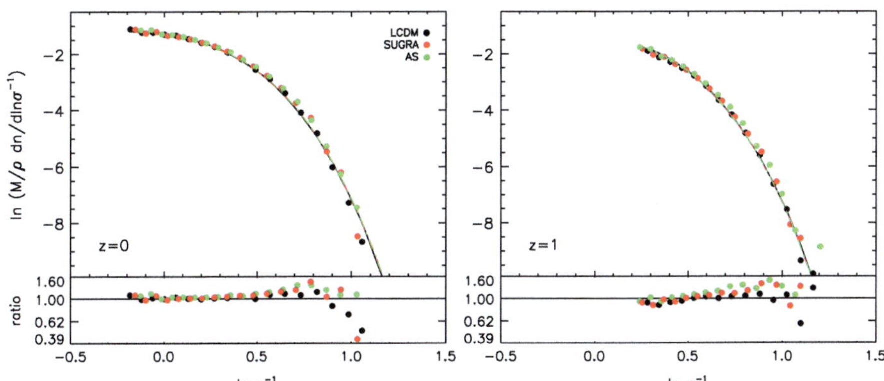

Fig. 3.14 The halo mass function for the SUGRA and AS model and ΛCDM at $z = 0$ and 1 compared to the Jenkins et al. (2001) analytic fit. The Jenkins et al. mass function is plotted as *solid black (red/green) lines* for ΛCDM (SUGRA/AS). Underneath each panel the ratio of the mass function measured from the simulation and the Jenkins et al. mass function is plotted for all models. Note a logarithmic scale is used on the y-axis in the ratio plots

in the spectral shape of the linear theory power spectrum from ΛCDM linear theory ($\Omega_m h^2 = 0.133$). As a result there are fewer low mass halos and a similar number of high mass haloes at $z = 0$ compared to ΛCDM. This change accounts for the decrease in the mass function for $M < 10^{15} h^{-1} M_\odot$ seen at $z = 0$ in the AS model (green squares). At $z = 0$, there are only small (<20 %) differences between the measured value and the Jenkins et al. fitting formula for $M < 10^{15} h^{-1} M_\odot$ for ΛCDM and the AS model from Stage III. The ratio between the Jenkins et al. mass function for ΛCDM and the measured mass function for ΛCDM and the AS model from Stage III is plotted underneath each panel in the second row in Fig. 3.13. Only the SUGRA and AS models are plotted in Fig. 3.13 but similar differences in halo abundances are seen in the INV models compared to ΛCDM, whilst only negligible differences with ΛCDM were found in the mass functions of 2EXP and CNR. Grossi and Springel (2009) found similar results for the mass function over the range 10^{11}–10^{14} $h^{-1} M_\odot$ in an 'early dark energy' model, using much smaller volume simulations than ours. They found a higher number density of haloes corresponding to groups and clusters in non-standard dark energy models at high redshifts compared to ΛCDM, while at $z = 0$ the models all agreed with one another. We find similar results although using the cosmological parameters from Table A.1 for each quintessence model can give different abundances at $z = 0$ in those models compared to ΛCDM because although σ_8 is the same the shape of the linear theory can be different. Also, we have been able to probe a higher mass range for the dark matter haloes. The high mass end of the mass function is very sensitive to changes in the current value of the linear growth factor in the different cosmologies.

In Fig. 3.14 we plot the fraction of the total mass in haloes of mass M rather than simply the abundance as shown in Fig. 3.13. We compare the Jenkins et al. analytic fit to our simulated halo mass functions in the SUGRA and AS models and in ΛCDM at $z = 0$ and 1 in Fig. 3.14. In this plot the quantity $\ln \sigma^{-1}(M, z)$ is used as the mass variable instead of M, where $\sigma^2(M, z)$ is the variance of the linear density field at $z = 0$. This variance can be expressed as

$$\sigma^2(M, z) = \frac{D^2(z)}{2\pi^2} \int_0^\infty k^2 P(k) W^2(k; M) dk \,, \tag{3.6}$$

where $W(k; M)$ is a top hat window function enclosing a mass M, $D(z)$ is the linear growth factor of perturbations at redshift z and $P(k)$ is the power spectrum of the linear density field. Plotting different masses at different redshifts in this way takes out the redshift dependence in the power spectrum. Note a large value of $\ln \sigma^{-1}(M, z)$ corresponds to a rare halo. Using this variable, Jenkins et al. found that the mass function at different epochs has a universal form, for a fixed power spectrum shape. Note that in our case, the Stage III simulations have somewhat different power spectra, which account for the bulk of the dispersion between the simulation results at the rare object end of Fig. 3.14; in Stage I, the simulation results agree with the Jenkins et al. universal form to within 25 % at $\ln \sigma^{-1} = 1.0$. As shown in Fig. 3.14, we find the Jenkins et al. fitting formula is accurate to $\sim$20 % at $z = 0$ for all the models in the range $M < 10^{15} h^{-1} M_\odot$. At higher redshifts the measured mass function for the SUGRA model and ΛCDM differ from the Jenkins et al. mass function by $\sim$30 % over the same mass range while for the AS model the difference is $\sim$50 % at $z = 1$. In previous work, Linder and Jenkins (2003) also found that the predicted mass function for a SUGRA-QCDM simulation, which would be the equivalent of our Stage I simulations, was well fit (within 20 %) by the Jenkins et al. formula.

3.3.5 The Appearance of Baryonic Acoustic Oscillations in Quintessence Models

In this section we examine the baryonic acoustic oscillation signal in the matter power spectrum for the AS, SUGRA and CNR models. Angulo et al. (2008) presented a detailed set of predictions for the appearance of the BAO signal in the ΛCDM model, covering the impact of nonlinear growth, peculiar velocities and scale dependent redshift space distortions and galaxy bias. Here we focus on the first of these effects and show power spectra in real space for the dark matter. We do not consider the INV1 model as it is not consistent with observational constraints (Appendix A.1), or the INV2 or 2EXP models as they are indistinguishable from ΛCDM, and hence were not simulated again in Stage III (Sect. 3.3.3).

In Stage I of our simulations (Sect. 3.3.1), we would expect the linear theory comoving BAO for the quintessence models to be identical to ΛCDM as the same linear theory power was used for all models. In Stage II (Sect. 3.3.2), some of the quintessence models have large amounts of dark energy at early times which will alter the sound horizon in these models compared to ΛCDM (see Table A.1), and as a result we would expect to see a corresponding shift in the BAO peak positions. The best fit cosmological parameters found in Stage III were derived using CMB, BAO and SN distance measurements (see Appendix A.1). Stage III of our simulations (Sect. 3.3.3) uses these parameters and we would expect models with the same BAO distance measures to have the same peak pattern in the matter power spectrum as ΛCDM.

The baryonic acoustic oscillations are approximately a standard ruler and depend on the sound horizon, r_s, given in Eq. A.1.3 (Sanchez et al. 2008). The apparent size of the BAO scale depends on the distance to the redshift of observation and on the ratio r_s/D_v, where D_v is an effective distance measure which is a combination of D_A and H, given in Eq. A.1.6. In most quintessence models, r_s remains unchanged unless there is appreciable dark energy at last scattering. Models which have the same ratio of r_s/D_v are impossible to distinguish using BAO.

To calculate the power spectrum for a galaxy redshift survey, the measured angular and radial separations of galaxies pairs are converted to co-moving separations and scales. This conversion is dependent on the cosmological model assumed in the analysis. These changes can be combined into the single effective measure, D_v. Once the power spectrum is calculated in one model we can simply re-scale $P(k)$ using D_v to obtain the power spectrum and BAO peak positions in another cosmological model (see Sánchez et al. 2009). In the left panel of Fig. 3.15, we plot the ratio of D_v in four quintessence models compared to ΛCDM up to $z = 1.5$. Percival et al. (2007) found $D_v = 564 \pm 23\,\mathrm{h}^{-1}\,\mathrm{Mpc}$ at $z = 0.2$ and $D_v = 1019 \pm 42\,\mathrm{h}^{-1}\,\mathrm{Mpc}$ at $z = 0.35$ using the observed scale of BAO measured from the SDSS DR5 galaxy sample and 2dFGRS. These data points are plotted as grey circles in Fig. 3.15. Note that at face value none of the models we consider are consistent with the Percival et al. (2007) point at $z = 0.35$. These authors report a 2.4σ discrepancy between their results using BAO and the constraints available at the time from supernovae. The blue square plotted in the left panel in Fig. 3.15 is the constraint $D_v = 1300 \pm 31\,\mathrm{Mpc}$ at $z = 0.35$ found by Sánchez et al. (2009). This constraint was found using a much larger LRG dataset and improved modelling of the correlation function on large scales. The constraint found by Sánchez et al. (2009) using CMB and BAO data is fully consistent with CMB and SN results. The results from Percival et al. (2010) for D_v and $r_s(z_d)/D_v$ at $z = 0.275$ using WMAP 5 year data together with the SDSS data release 7 galaxy samples are also plotted (black triangles). The Percival et al. (2010) results are in much better agreement with those of Sánchez et al. (2009).

Over the range of redshifts plotted in Fig. 3.15 the distance measure, D_v, in the AS, 2EXP and CNR models differ from ΛCDM by at most 2 % and is <1 % in these models for $z < 0.2$. Re-scaling the power spectrum for these dark energy cosmologies would result in a small shift ~ 1 % in the position of the peaks at low redshifts. The value of D_v in the SUGRA model differs from ΛCDM by at most 9 % up to $z = 1.5$. The right panel in Fig. 3.15 shows the ratio of $r_s(z_d)/D_v$ in the quintessence models compared to ΛCDM, where r_s is the co-moving sound horizon scale at the drag redshift, z_d, which we discuss in Appendix A.1. The value of $r_s(z_d)/D_v$ can be constrained using the position of the BAO in the power spectrum.

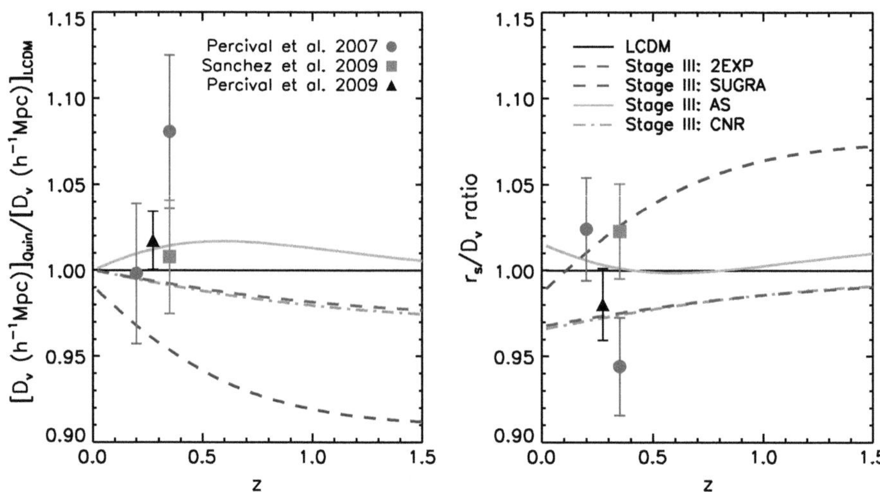

Fig. 3.15 The ratio of the distance measure $D_v(z)$ (*left panel*) and the ratio of $r_s(z_d)/D_v$ (*right panel*) for four quintessence models compared to ΛCDM as indicated by the key in the *right hand panel*. The *grey circles* are estimate points from Percival et al. (2007) at $z = 0.2$ and $z = 0.35$ measured using the observed scale of BAO calculated from the SDSS and 2dFGRS main galaxy samples. Sánchez et al. (2009) combined CMB data with information on the shape of the redshift space correlation function using a larger LRG dataset and found $D_v(z = 0.35) = 1300 \pm 31$ Mpc and $r_s(z_d)/D_v = 0.1185 \pm 0.0032$ at $z = 0.35$ (*blue squares*). The data points from Percival et al. (2010) for D_v and $r_s(z_d)/D_v$ at $z = 0.275$ using WMAP 5 year data + SDSS DR7 are plotted as *black triangles* (colour figure online)

In the right panel of Fig. 3.15 the grey symbols are the results from Percival et al. (2007) at $z = 0.2$ and $z = 0.35$. From this plot it is clear that the SUGRA and AS model are within the 1σ limits at $z = 0.2$. The 2EXP and CNR model lie just outside the 1σ errors at $z = 0.35$. Note the value of $r_s(z_d)/D_v$ for ΛCDM at $z = 0.35$ also lie outside the 1σ errors, see Percival et al. (2010) for more detail. The blue square plotted in the right panel in Fig. 3.15 is $r_s(z_d)/D_v = 0.1185 \pm 0.0032$ at $z = 0.35$ and was obtained using information on the redshift space correlation function together with CMB data (Sánchez et al. 2009).

In Figs. 3.16 and 3.17 we plot the $z = 0$ and $z = 3$ power spectra in the AS and SUGRA models divided by a linear theory ΛCDM reference spectrum which has been smoothed using the coarse rebinning method proposed by Percival et al. (2007) and refined by Angulo et al. (2008). After dividing by this smoothed power spectrum, the acoustic peaks are more visible in the quasi-linear regime. In Figs. 3.16 and 3.17, the measured power in each bin has been multiplied by a factor, f, to remove the scatter due to the small number of large scale modes in the simulation (Baugh and Efstathiou 1994; Springel et al. 2005). This factor, $f = P(k)_{\text{linear}}/P(k)_{\text{N-body}}$, is the ratio of the expected linear theory power and the measured power in each bin at $z = 5$, at which time the power on these scales is still expected to be linear.

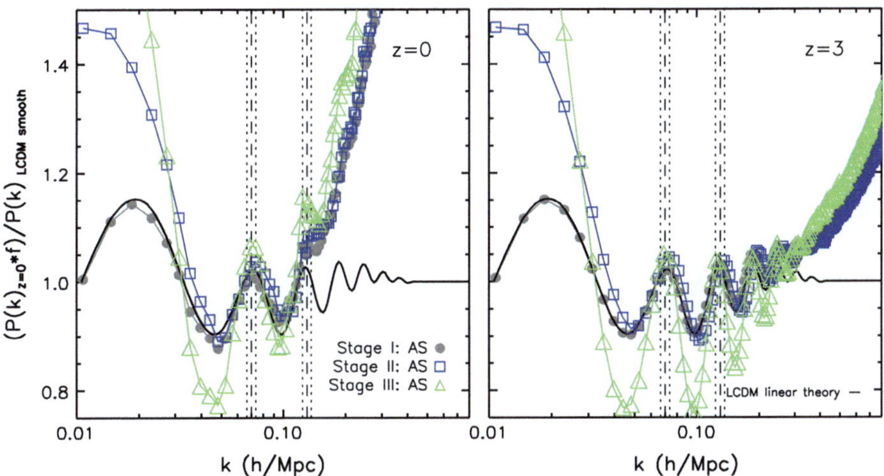

Fig. 3.16 The real space power spectrum for the AS model on large scales at $z = 0$ (*left*) and $z = 3$ (*right*). All power spectra have been divided by a smoothed linear 'no-wiggle' theory $P(k)$ for ΛCDM. The factor, f, removes the scatter of the power measured in the simulation around the expected linear theory power. Stage I in our simulation is represented by *grey circles*, Stage II is represented by *open blue squares* and Stage III results are shown as *green triangles*. The *black solid line* represents the linear theory power spectrum in ΛCDM divided by the smooth reference spectrum. The *vertical dashed* (*dotted*) *lines* show the position of the first two acoustic peaks (positions $\pm 5\,\%$) for a ΛCDM cosmology

Multiplying by this correction factor allows us to see the onset of nonlinear growth around $k \sim 0.15\,\mathrm{h\,Mpc^{-1}}$ more clearly.

In Fig. 3.16 (3.17) we plot the AS (SUGRA) power spectrum as grey circles from Stage I, blue (purple) squares from Stage II and green (red) triangles from Stage III. The black line represents the linear theory power in ΛCDM divided by the smooth reference spectrum. In both plots and for all power spectra, the same reference spectrum is used. The reference is a simple 'wiggle-free' CDM spectrum, with a form controlled by the shape parameter $\Gamma = \Omega_{\mathrm{m}} h$ (Bardeen et al. 1986). The difference between the AS and ΛCDM linear theory, as shown in Fig. 3.11, results in an increase in large scale power on scales $k < 0.04\,\mathrm{h\,Mpc^{-1}}$. The vertical dashed (dotted) lines show the first two positions of the acoustic peaks (positions $\pm 5\,\%$) for a ΛCDM cosmology.

As shown in Fig. 3.16, we find that the position of the first acoustic peak in the AS model from Stage I is the same as in ΛCDM. The position of the first peak for the AS model, measured in Stage II of our simulations (blue squares), is slightly shifted ($\sim 4\,\%$) to smaller scales compared to ΛCDM as the sound horizon is altered in the AS model. In Stage III, when the best fit cosmological parameters for the AS model are used, the sound horizon in the AS model and ΛCDM are very similar at $z \sim 1090$ and there is a very small ($<1\,\%$) shift in the position of the first peak

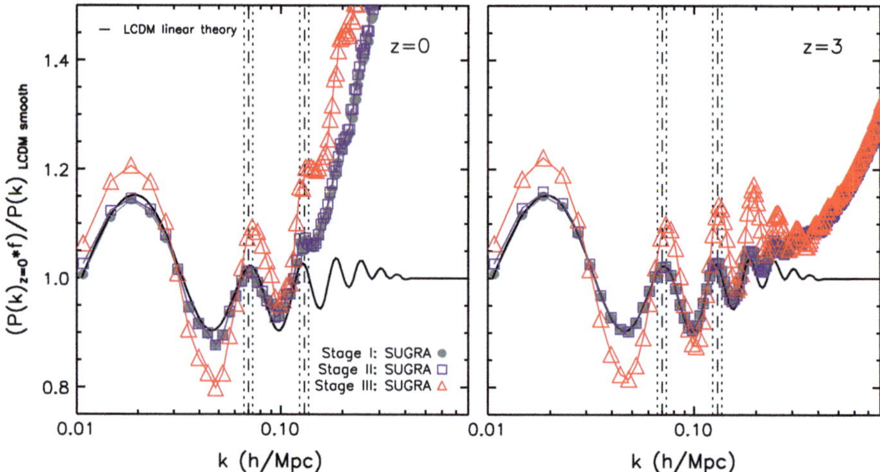

Fig. 3.17 The real space power spectrum for the SUGRA model on large scales at $z = 0$ and $z = 3$. All power spectra have been divided by a smoothed linear theory $P(k)$ for ΛCDM. Stage I in our simulation is represented by *grey circles*, Stage II is represented by *open purple squares* and Stage III results are shown as *red triangles*. The *black solid line* represent the linear theory power spectrum in ΛCDM divided by the smooth reference spectrum. The *vertical dashed (dotted) lines* show the position of the first two acoustic peaks (positions $\pm 5\,\%$) for a ΛCDM cosmology

(green triangles). As there is less nonlinear growth at $z = 3$ the higher order peaks are more visible in the right-hand plot in Fig. 3.16.

In Fig. 3.17, the SUGRA power spectrum from Stages I, II and III are plotted. The SUGRA $P(k)$ from Stages I and II have identical peak positions to ΛCDM as the sound horizon is the same as in ΛCDM in these cases. There is a shift ($\sim 5\,\%$) in the position of the first peak in the SUGRA model using the $P(k)$ measured in Stage III. Note the units on the x-axis are h/Mpc and from Table A.1, $h = 0.67$ for the Stage III SUGRA model compared to $h = 0.715$ for ΛCDM. On small scales the BAO signature is damped due to more nonlinear structure formation at $z = 0$ compared to $z = 3$ as shown in Fig. 3.17. We find a large increase in the power in the region of the second peak, $k \sim 0.15\,\mathrm{h\,Mpc}^{-1}$ in both the AS and SUGRA models, measured in Stage III, compared to ΛCDM. For brevity we have not included the plots of the power spectra for the CNR model showing the baryonic acoustic oscillations. We find identical peak positions in ΛCDM and this model in all stages at $z = 0$.

The AS and SUGRA model are very different to ΛCDM at late times and as result they affect the growth of structure at $z > 0$ as seen in Sects. 3.3.3 and 3.3.4. We have found that models like this do not necessarily have different BAO peak positions to ΛCDM in the matter power spectrum. These results suggest that distinguishing a quintessence model, like the AS model used in this chapter, using measurements of the BAO peak positions in future galaxy surveys, will be extremely difficult. The BAO peak positions for the CNR model will be shifted by at most $2\,\%$ in the range $z < 1.5$ compared to ΛCDM after re-scaling the power spectra by D_V. In conclusion it is possible to have quintessence cosmologies with higher levels of dark energy at early times than in ΛCDM and still measure the same peak positions for the BAO in the matter power spectrum.

3.4 Summary

Observing the dynamics of dark energy is the central goal of future galaxy surveys and would distinguish a cosmological constant from a dynamical quintessence model. Using a broad range of quintessence models, with either a slowly or rapidly varying equation of state, we have analysed the influence of dynamical dark energy on structure formation using N-body simulations.

We have considered a range of quintessence models that can be classified as either 'tracking' models, for example the SUGRA and INV models, or 'scaling' solutions, such as the AS, CNR or 2EXP models, depending on the evolution of their equation of state (see Table 3.1; Sect. 3.2). The models feature both rapidly and slowly varying equations of state and the majority of the models could be classified as 'early dark energy' models as they have a non-negligible amount of dark energy at early times.

In order to accurately mimic the dynamics of the original quintessence models at high and low redshift, it is necessary to use a general prescription for the dark energy equation of state which has more parameters than the ubiquitous 2 variable equation. Parametrizations for w which use 2 variables are unable to faithfully represent dynamical dark energy models over a wide range of redshifts and can lead to biases when used to constrain parameters (Bassett et al. 2004). Our task has been made easier by the availability of parametrizations which accurately describe the dynamics of the different quintessence models (Corasaniti and Copeland 2003; Linder and Huterer 2005). This allows us to modify the Friedmann equation in the simulation, using the equation of state as a function of redshift. We use the parametrization of Corasaniti and Copeland (2003). In its full six parameter form, this framework can describe the quintessence model back to the epoch of nucleosynthesis. Four parameters are sufficient to describe the behaviour of the quintessence field over the redshift interval followed by the simulations. With this description of the equation of state, our simulations are able to accurately describe the impact of the quintessence model on the expansion rate of the Universe, from the starting redshift to the present day. This would not be the case with a 2 parameter model for the equation of state.

In this thesis we have taken into account three levels of modification from a ΛCDM cosmology which are necessary if we wish to faithfully incorporate the effects of quintessence dark energy into a N-body simulation. The first stage is to replace the cosmological constant with the quintessence model in the Friedmann equation. A quintessence model with a different equation of state from $w = -1$ will lead to a universe with a different expansion history. This in turn alters the rate at which perturbations can collapse under gravity. The second stage is to allow the change in the expansion history and perturbations in the quintessence field to have an impact on the form of the linear theory power spectrum. The shape of the power spectrum can differ significantly from ΛCDM on large scales if there is a non-negligible amount of dark energy present at early times. This alters the shape of the turn-over in the power spectrum compared to ΛCDM. Thirdly, as the quintessence model should be consistent with observational constraints, the cosmological parameters used for the dark energy model could be different from the best fit ΛCDM parameters. In the three stages of simulations we look at the effect each of the above modifications has on the nonlinear growth of structure. Deconstructing the simulations into three stages allows us to isolate specific features in the quintessence models which play a key role in the growth of dark matter perturbations. In the first stage of comparison, in which all that is changed is the expansion history of the universe, we found that some of the quintessence models showed enhanced structure formation at $z > 0$ compared to ΛCDM. The INV1, INV2, SUGRA and AS models have slower growth rates than ΛCDM. Hence, when normalising to the same σ_8 today, structures must form at earlier times in these models to overcome the lack of growth at late times. Models such as the 2EXP and CNR model have the same recent growth rate as ΛCDM and showed no difference in the growth of structure. The difference in linear and nonlinear growth can largely be explained by the difference in the growth factor at different epochs in the models. At the same growth factor, the power in the models only diverges at the 15 % level well into the nonlinear regime.

In the second stage, a self-consistent linear theory $P(k)$ was used for each quintessence model to generate the initial conditions in the simulations. The amount of dark energy present at early times will determine the impact on the linear theory dark matter power spectrum and the magnitude of deviation from the ΛCDM spectrum. High levels of dark energy at early times suppress the growth of the dark matter on scales inside the horizon, resulting in a broader turn-over in the power spectrum. We found that models with the highest levels of dark energy at the last scattering surface, such as the AS and CNR models, have linear theory $P(k)$ which differ the most from ΛCDM. The results of the N-body simulations of the AS and the SUGRA model show a very small increase in nonlinear growth compared to the results in Stage I. The increase in the linear theory power is on very large scales and does not change the small scale growth significantly.

In our final stage of simulating the effects of quintessence, we found the best fitting cosmological parameters for each model, $\Omega_m h^2$, $\Omega_b h^2$ and H_0, consistent with current CMB, SN and BAO measurements. For quintessence dark energy models, it is important to consider the changes in more than just one cosmological parameter when fitting to the observational data. For example, for a given dark energy equation of state, the values of $\Omega_m h^2$ and H_0 may change in such a way to compensate one another and give similar growth rates and expansion histories to ΛCDM. These compensating effects will be missed if, for example, only Ω_m is changed for the dark energy model as in recent work (Alimi et al. 2010). Models with cosmological parameters which fit the data but were significantly different from ΛCDM were simulated again (Sect. 3.3.3).

We will now summarise and discuss the main results for each model. The key features of each of the quintessence models are presented in Table 3.2. The INV1 model was unable to fit the data with a reasonable χ^2/ν (Table A.1). This toy model

Table 3.2 The key features in the evolution of the quintessence models simulated

Model	Transition type	Transition redshift	$\Omega_{\mathrm{DE}}(z = 300)$	$\Delta D(z = 5)$ (%)
INV1	Gradual	~4.5	~0.009	~50
INV2	Gradual	~5	Negligible	~10
SUGRA	Rapid	~9	~0.01	~20
2EXP	Rapid	~4	~ 0.015	0
CNR	Rapid	~5.5	~0.03	0
AS	Rapid	~1	~0.11	20

$\Delta D(z = 5)$ is the ratio of the linear growth factor for each quintessence model compared to ΛCDM at $z = 5$. A late time transition in the equation of state is defined as occurring at $z < 2$. The AS, CNR, 2EXP and SUGRA models can be considered as 'early dark energy' models as they have non-negligible amounts of dark energy present at early times

had the largest growth factor ratio to ΛCDM at $z = 5$ and as a result showed the most enhanced growth in Stage I of our simulations. The linear growth factor for the INV2 model is very different to ΛCDM at early times and gives rise to enhanced growth at $z > 0$ as seen in Sect. 3.3.1. This model has negligible dark energy at early times and so the spectral shape is not altered in Stage II. In the 2EXP model the rapid transition to $w = -1$ in the equation of state early on leaves little impact on the growth of dark matter and as a result the power spectra and mass function are indistinguishable from ΛCDM. As both the INV2 and 2EXP models already agree with cosmological measurements with very similar values for $\Omega_{\mathrm{m}}h^2$, $\Omega_{\mathrm{b}}h^2$ and H_0 to ΛCDM, we did not run these simulations again. The SUGRA model has enhanced linear and nonlinear growth and halo abundances compared to ΛCDM at $z > 0$ and an altered linear theory power spectrum shape. The mass function results for all stages of our simulations for the SUGRA model show enhanced halo abundances at $z > 0$. Analysing the SUGRA power spectra, from a Stage III simulation which used the best fit parameters for this model, reveals a ~5 % shift in the position of the first BAO peak. We find the distance measure D_v for the SUGRA model differs by up to 9 % compared to ΛCDM over the range $0 < z < 1.5$. Re-scaling the power measured for the SUGRA model by the difference in D_v would result in an even larger shift in the position of the BAO peaks.

The CNR model has high levels of dark energy early on which alters the spectral shape on such large scales that the nonlinear growth of structure is only slightly less than ΛCDM at $z < 5$. This model has a halo mass abundance at $z < 5$ and BAO peak positions at $z = 0$ which are the same as in ΛCDM. For $z < 0.5$ the distance measure, D_v, for the CNR model differs from ΛCDM by ~1 %, as result there would be a corresponding small shift in the BAO peak positions. The rapid early transition at $z = 5.5$ in the equation of state to $w_0 = -1$ in this model seems to remove any signal of the large amounts of dark energy at early times that might be present in the growth of dark matter perturbations.

The AS model has the highest levels of dark energy at early times, and so its linear theory spectrum is altered the most. This results in a large increase in large scale power, when we normalise the power spectrum to $\sigma_8 = 0.8$ today. The results from Stage III using the best fit parameters show both enhanced linear and nonlinear growth at $z < 5$. The linear theory $P(k)$ is altered on scales $k \sim 0.1\,h\,\mathrm{Mpc}^{-1}$ which drives an increase in nonlinear growth on small scales compared to ΛCDM. The mass function results in Stage III for this model show enhanced halo abundances at $z > 0$. We find that using the best fit cosmological parameters for the AS model produces a BAO profile with peak positions similar to those in ΛCDM. At low redshifts there is $\sim 1\,\%$ shift in the first peak compared to ΛCDM after re-scaling the power with the difference in the distance measure D_v between the two cosmologies.

These results from Stage III of our N-body simulations show that dynamical dark energy models in which the dark energy equation of state makes a late $(z < 2)$ rapid transition to $w_0 = -1$ show enhanced linear and nonlinear growth compared to ΛCDM at $z > 0$ and have a greater abundance of dark matter haloes compared to ΛCDM for $z > 0$. We found that dynamical dark energy models can be significantly different from ΛCDM at late times and still produce similar BAO peak positions in the matter power spectrum. Models which have a rapid early transition in their dark energy equation of state and mimic ΛCDM after the transition, show the same linear and nonlinear growth and halo abundance as ΛCDM for all redshifts. We have found that these models can give rise to BAO peak positions in the matter power spectrum which are the same as those in a ΛCDM cosmology. This is true despite these models having non-negligible amounts of dark energy present at early times.

References

Albrecht AJ, Skordis C (2000) Phys Rev Lett 84:2076
Alimi JM, Fuzfa A, Boucher V, Rasera Y, Courtin J, Corasaniti PS (2010) MNRAS 771:405
Angulo R, Baugh CM, Frenk CS, Lacey CG (2008) MNRAS 383:755
Bardeen JM, Bond JR, Kaiser N, Szalay AS (1986) ApJ 304:15
Barreiro T, Copeland EJ, Nunes NJ (2000) Phys Rev D 61:127301
Bassett BA, Kunz M, Silk J, Ungarelli C (2002) MNRAS 336:1217
Bassett BA, Corasaniti PS, Kunz M (2004) ApJ 617:L1
Baugh CM, Efstathiou G (1994) MNRAS 270:183
Baumgart DJ, Fry JN (1991) ApJ 375:25
Bean R, Hansen SH, Melchiorri A (2001) Phys Rev D 64:103508
Brax P, Martin J (1999) Phys Lett B 468:40
Caldwell RR, Doran M, Mueller CM, Schafer G, Wetterich C (2003) ApJ 591:L75
Copeland EJ, Nunes NJ, Rosati F (2000) Phys Rev D 62:123503
Corasaniti PS (2004) astro-ph/0401517
Corasaniti PS, Copeland EJ (2003) Phys Rev D 67:063521
Corasaniti PS, Kunz M, Parkinson D, Copeland EJ, Bassett BA (2004) Phys Rev D 70:083006
Crocce M, Fosalba P, Castander FJ, Gaztanaga E (2010) MNRAS 403:1353
Doran M, Robbers G (2006) JCAP 0606:026
Doran M, Robbers G, Wetterich C (2007) Phys Rev D 75:023003
Efstathiou G, Rees MJ (1988) MNRAS 230:5P
Eke VR, Cole S, Frenk CS (1996) MNRAS 282:263
Eke VR, Navarro JF, Steinmetz M (2001) ApJ 554:114
Fang W, Hu W, Lewis A (2008) Phys Rev D 78:087303
Ferreira PG, Joyce M (1998) Phys Rev D 58:023503

Francis MJ, Lewis GF, Linder EV (2008) MNRAS 394:605
Gerke BF, Efstathiou G (2002) MNRAS 335:33
Governato F et al (1999) MNRAS 307:949
Grossi M, Springel V (2009) MNRAS 394:1559
Halliwell JJ (1987) Phys Lett B 185:341
Hockney RW, Eastwood JW (1981) Computer simulation using particles. McGraw-Hill, New York
Jenkins A, Frenk CS, White SDM, Colberg JM, Cole S, Evrard AE, Couchman HMP, Yoshida N (2001) MNRAS 321:372
Kunz M, Corasaniti P-S, Parkinson D, Copeland EJ (2004) Phys Rev D 70:041301
Lacey CG, Cole S (1994) MNRAS 271:676
Lewis A, Bridle S (2002) Phys Rev D 66:103511
Linder EV, Huterer D (2005) Phys Rev D 72:043509
Linder EV, Jenkins A (2003) MNRAS 346:573
Percival WJ et al (2007) MNRAS 381:1053
Percival WJ et al (2010) MNRAS 401:2148
Press WH, Schechter P (1974) ApJ 187:425
Reed D, Bower R, Frenk C, Jenkins A, Theuns T (2007) MNRAS 374:2
Sanchez AG, Baugh CM, Angulo R (2008) MNRAS 390:1470
Sánchez AG, Crocce M, Cabré A, Baugh CM, Gaztañaga E (2009) MNRAS 400:1643
Sheth RK, Tormen G (2002) MNRAS 329:61
Sheth RK, Mo HJ, Tormen G (2001) MNRAS 323:1
Smith RE et al (2003) MNRAS 341:1311
Springel V et al (2005) Nature 435:629
Steinhardt PJ, Wang L-M, Zlatev I (1999) Phys Rev D 59:123504
Wands D, Copeland EJ, Liddle AR (1993) NYASA 688:647
Warren MS, Abazajian K, Holz DE, Teodoro L (2006) ApJ 646:881
Weller J, Lewis AM (2003) MNRAS 346:987
Wetterich C (1995) Astron Astrophys 301:321
Wetterich C (2004) Phys Lett B 594:17
White SDM, Efstathiou G, Frenk CS (1993) MNRAS 262:1023
Xia J-Q, Viel M (2009) JCAP 0904:002
Zlatev I, Wang L-M, Steinhardt PJ (1999) Phys Rev Lett 82:896

Chapter 4
Modelling Redshift Space Distortions in Hierarchical Cosmologies

4.1 Introduction

Galaxy redshift surveys allow us to study the 3D spatial distribution of galaxies and clusters. In a homogeneous universe, redshift measurements would probe only the Hubble flow and would provide accurate radial distances for galaxies. In reality, peculiar velocities are gravitationally induced by inhomogeneous structure and distort the measured distances. Kaiser (1987) described the anisotropy of the clustering pattern in redshift space but restricted his calculation to large scales where linear perturbation theory should be applicable. In the linear regime, the matter power spectrum in redshift space is a function of the power spectrum in real space and the parameter $\beta = f/b$ where f is the linear growth rate. The linear bias factor, b, characterises the clustering of galaxies with respect to the underlying mass distribution (e.g. Kaiser 1987). Scoccimarro (2004) extended the analysis of Kaiser (1987) into the non-linear regime, including the contribution of peculiar velocities on small scales. We study the distortions in the redshift space power spectrum in ΛCDM and quintessence dark energy models, using large volume N-body simulations, and test predictions for the form of the redshift space distortions.

In previous work, Cole et al. (1994) and Hatton and Cole (1998) examined the linear approximations made by Kaiser (1987) and showed that non-linearities in the velocity and density perturbations affect the anisotropy of the redshift space power spectrum out to surprisingly large scales. Using N-body simulations in a periodic cube of $300\,h^{-1}$Mpc on a side, Cole et al. (1994) found that the measured value of β deviates from the Kaiser formula on wavelengths of $50\,h^{-1}$ Mpc or more as a result of these non-linearities. Hatton and Cole (1998) extended this analysis to slightly larger scales using the Zel'dovich approximation combined with a dispersion model where non-linear velocities are treated as random perturbations to the linear theory velocity. In both these studies, the scales at which a departure from linear theory was seen pushed the simulation results to the very limit. Velocity perturbations converge more slowly than density perturbations, and so very large computational boxes are essential for accurate predictions. These previous studies do not provide an accurate

E. Jennings, *Simulations of Dark Energy Cosmologies*, Springer Theses, DOI: 10.1007/978-3-642-29339-9_4, © Springer-Verlag Berlin Heidelberg 2012

description of the non-linearities in the velocity field as the Zel'dovich approximation does not model the velocities correctly, as it only treats part of the bulk motions, and in a computational box of length $300\,h^{-1}$Mpc, the power which determines the bulk flows has not converged. Scoccimarro (2004) measured the large scale form of the redshift space power spectrum using the VLS simulation of the Virgo consortium in a box of length $479\,h^{-1}$Mpc (Yoshida et al. 2001), and found discrepancies from the Kaiser formula on scales $k > 0.1\,h$Mpc^{-1}. Assuming a ΛCDM cosmology, Scoccimarro (2004) also found significant non-linear corrections due to the evolution of the velocity fields on large scales. In this chapter, we focus on the impact of non-linearities and determine their impact on the redshift space power spectrum in ΛCDM and quintessence dark energy models. The volume of our simulations, detailed in Chaps. 2 and 3, is 125 times larger than that used by Cole et al. (1994) and approximately 27 times larger than the one used by Scoccimarro (2004), and allow us to accurately predict the redshift space distortions for each cosmology out to very large scales.

Percival and White (2009) investigated the redshift space clustering using a N-body simulation in a $1\,h^{-1}$Gpc box. They argued that large scale redshift space distortions can provide a bias independent constraint on $f\sigma_8$(mass). By decomposing the redshift space power spectrum into multipole moments, Percival and White (2009) then fitted to the measured monopole moment of the power spectrum to extract the galaxy-galaxy and velocity-velocity power spectra. In this thesis we do not address the issue of bias and we measure the velocity power spectra directly from the simulations to test deviations from linear perturbation theory.

This chapter is organised as follows: In Sect. 4.2 we discuss the linear growth rate and review the theory of redshift space distortions on linear and non-linear scales. The quintessence models considered in this chapter have already been discussed in Chap. 3. The main results of this chapter are presented in Sects. 4.3 and 4.4. The linear theory redshift space distortion, as well as models for the redshift space power spectrum which include non-linear effects are examined in Sect. 4.3 for various dark energy cosmologies. In Sect. 4.4 we present the density-velocity relation measured from the simulations. Using this relation the non-linear models used in the previous section can be made cosmology independent. We present a prescription for obtaining the non-linear velocity divergence power spectrum from the non-linear matter power spectrum at an arbitrary redshift in Sect. 4.4.2. Our summary are presented in Sect. 4.5.

4.2 Redshift Space Distortions

In Sect. 4.2.1 we consider several parametrizations which are commonly used for the linear growth rate. In Sect. 4.2.2 we review linear perturbation theory for redshift space distortions and discuss the assumptions that are used in this approach. In Sect. 4.2.3 we present several models proposed to describe the distortions in the non-linear regime. A similar review can be found in Percival and White (2009).

4.2.1 Linear Growth Rate as a Probe of Gravity

The linear growth rate is a promising probe of the nature of dark energy (Guzzo et al. 2008; Wang 2008; Linder 2008; Song and Percival 2009; White et al. 2009; Percival and White 2009; Stril et al. 2009; Simpson and Peacock 2010). Although the growth equation for dark matter perturbations is easy to solve exactly, it is common to consider parametrizations for the linear growth rate, $f = \mathrm{d}\ln D/\mathrm{d}\ln a$, where $D(a)$ is the linear growth factor, see Chap. 2. These parametrizations employ different variables with distinct dependencies on the expansion and growth histories.

A widely used approximation for f, first suggested by Peebles (1976), is $f(z) \approx \Omega_\mathrm{m}^{0.6}$. Lahav et al. (1991) found an expression for f, in terms of the present day densities of matter, Ω_m, and dark energy, Ω_DE, which showed only a weak dependence on the dark energy density, with $f \approx \Omega_\mathrm{m}^{0.6} + \Omega_\mathrm{DE}/70\,(1 + \Omega_\mathrm{m}/2)$. Linder (2005) extended the analysis of Wang and Steinhardt (1998) to find a new fitting formula to the exact solution for the growth factor, which he cast in the following form

$$g(a) = \frac{D(a)}{a} \approx \exp\left(\int_0^a \mathrm{d}\ln a\,[\Omega_\mathrm{m}^\gamma(a) - 1]\right), \tag{4.1}$$

where γ is the index which parametrises the growth history, while the expansion history is described by the matter density $\Omega_\mathrm{m}(a)$. (Linder 2005) proposed the empirical result $\gamma = 0.55 + 0.05[1 + w(z = 1)]$, where w is the dark energy equation of state, which gives $f = \Omega_\mathrm{m}^{0.55}$ for a cosmological constant (see also Linder and Cahn 2007).

In this chapter we consider three quintessence models, each with a different evolution for the dark energy equation of state parameter, $w(a)$. These models are a representative sample of a range of quintessence models and are a subset of those considered in Chap. 3, namely the SUGRA, the 2EXP and the CNR quintessence model. In the left panel of Fig. 4.1, we plot the exact solution for the linear theory growth factor, divided by the scale factor, as a function of redshift together with the fitting formula in Eq. 4.1. The 2EXP quintessence model is not plotted in Fig. 4.1 as the linear growth factor for this model differs from ΛCDM only at high redshifts, $z > 10$. Linder (2005) found that the formula in Eq. 4.1 reproduces the growth factor to better than 0.05 % for ΛCDM cosmologies and to $\sim$0.25 % for different dynamical quintessence models to the ones considered in this chapter. We have verified that this fitting formula for D is accurate to $\sim$1 % for the SUGRA and 2EXP dark energy models used in this chapter, over a range of redshifts. Note, in cosmological models which feature non negligible amounts of dark energy at high redshifts, a further correction factor is needed to this parametrisation (Linder 2009). Using the parametrization for $w(a)$ provided by Doran and Robbers (2006) for 'early dark energy', Linder (2009) proposed a single correction factor which was independent of redshift. The CNR model has a high fractional dark energy density at early times and as a result we do not expect the linear theory growth to be accurately reproduced by Eq. 4.1. As can be seen in Fig. 4.1 for the CNR model, any correction factor between the fitting formula suggested by Linder (2005) and the exact solution for

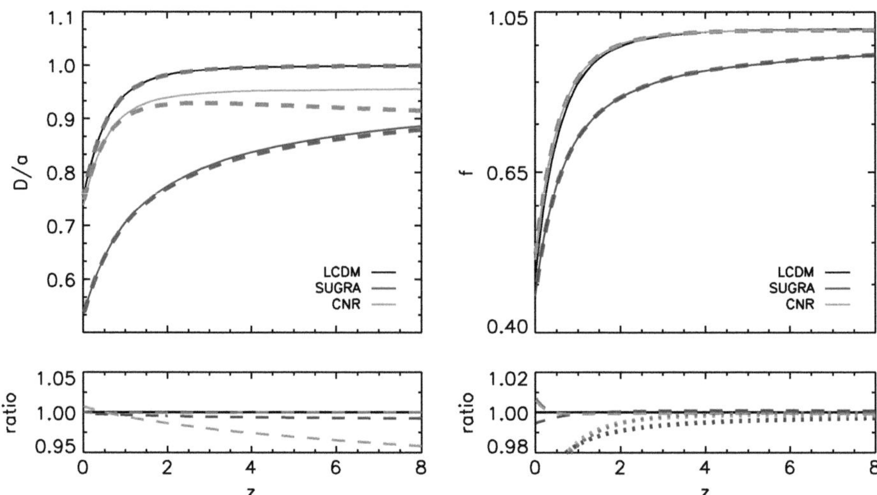

Fig. 4.1 *Left panel* the linear growth factor divided by the scale factor as a function of redshift for the SUGRA and CNR quintessence models and ΛCDM, as indicated by the key. *Right panel* the linear growth rate, $f = \mathrm{d}\ln D/\mathrm{d}\ln a$, for the two dark energy models and ΛCDM as a function of redshift. In both the *left* and *right main panels*, *solid lines* represent the exact solution for the linear growth factor and growth rate and *dashed lines* show the fitting formula given in Eq. 4.1. Note in the *right main panel* the ΛCDM *grey dashed line* has been omitted for clarity. The *lower left hand panel* shows the formula for $D(a)/a$ given by Linder (2005) divided by the exact solution as a function of redshift. The ratio of the formula in Eq. 4.1 for the growth rate, f, to the exact solution is shown in the *lower right hand panel*. Also in the *lower right panel* the *dotted lines* show the ratio of the fitting formula $f = \Omega_{\mathrm{m}}^{0.6}$ to the exact solution for each of the dark energy models plotted as a function of redshift

D/a would depend on redshift and is not simply a constant. In this case, the 'early dark energy' parametrisation of Doran and Robbers (2006) is not accurate enough to fully describe the dynamics of the CNR quintessence model. This difference is $\sim 5\,\%$ at $z = 8$ for the CNR model, as can be seen in the ratio plot in the left panel of Fig. 4.1. The exact solution for the linear growth rate, f, and the fitting formula in Eq. 4.1, $f = \Omega_{\mathrm{m}}^{\gamma}(a)$, is plotted in the right panel of Fig. 4.1. The old approximation $f = \Omega_{\mathrm{m}}^{0.6}$, is plotted in the bottom right panel in Fig. 4.1. The dotted lines represent the ratio $f = \Omega_{\mathrm{m}}^{0.6}$ to the exact solution for each of the dark energy models. It is clear that this approximation for the growth factor is not as accurate as the formula in Eq. 4.1 over the same range of redshifts.

4.2.2 Linear Redshift Space Distortions

The comoving distance to a galaxy, $\vec{s}$, differs from its true distance, $\vec{x}$, due to its peculiar velocity, $\vec{v}(\vec{x})$ (i.e. an additional velocity to the Hubble flow), as

$$s = x + \frac{\vec{v} \cdot \hat{x}}{H(a)}, \tag{4.2}$$

where $H(a)$ is the Hubble parameter and $\vec{v}\cdot\hat{x}$ is the peculiar velocity along the line of sight. Inhomogeneous structure in the universe induces peculiar motions which distorts the clustering pattern measured in redshift space on all scales. This effect must be taken into account when analyzing three dimensional datasets which use redshift as the radial coordinate. Redshift space effects alter the appearance of the clustering of matter, and together with non-linear evolution and bias, lead the power spectrum to depart from simple linear perturbation theory predictions.

On small scales, randomised velocities associated with viralised structures decrease the power. The dense central regions of galaxy clusters look elongated along the line of sight in redshift space, which produces 'fingers of God' (Jackson 1972) in redshift survey cone plots. On large scales, coherent bulk flows distort clustering statistics (see Hamilton 1998 for a review of redshift space distortions). For growing perturbations on large scales, the overall effect of redshift space distortions is to enhance the clustering amplitude. Any difference in the velocity field due to mass flowing from underdense regions to high density regions will alter the volume element, causing an enhancement of the apparent density contrast in redshift space, $\delta_s(\vec{r})$, compared to that in real space, $\delta_r(\vec{r})$. This effect was first analyzed by Kaiser (1987) and can be approximated by

$$\delta_s(r) = \delta_r(r)(1 + \mu^2 \beta), \tag{4.3}$$

where μ is the cosine of the angle between the wavevector, $\vec{k}$, and the line of sight, $\beta = f/b$ and the bias, $b = 1$ for dark matter.

The Kaiser formula (Eq. 4.3) relates the overdensity in redshift space to the corresponding value in real space using several approximations:

1. The small scale velocity dispersion can be neglected.
2. The velocity gradient $|d\vec{u}/dr| \ll 1$.
3. The velocity and density perturbations satisfy the linear continuity equation.
4. The real space density perturbation is assumed to be small, $|\delta(r)| \ll 1$, so that higher order terms can be neglected.

All of these assumptions are valid on scales that are well within the linear regime and will break down on different scales as the density fluctuations grow. The linear regime is therefore defined over a different range of scales for each effect.

The matter power spectrum in redshift space can be decomposed into multipole moments using Legendre polynomials, $L_l(\mu)$,

$$P(k, \mu) = \sum_{l=0}^{2l} P_l(k) L_l(\mu). \tag{4.4}$$

The anisotropy in $P(\vec{k})$ is symmetric in μ, as $P(k, \mu) = P(k, -\mu)$, so only even values of l are summed over. Each multipole moment is given by

$$P_l^s(k) = \frac{2l+1}{2} \int_{-1}^{1} P(k, \mu) L_l(\mu) \mathrm{d}\mu, \tag{4.5}$$

where the first two non-zero moments have Legendre polynomials, $L_0(\mu) = 1$ and $L_2(\mu) = (3\mu^2 - 1)/2$. Using the redshift space density contrast, Eq. 4.3 can be used to form $P(k, \mu)$ and then integrating over the cosine of the angle μ gives the spherically averaged monopole power spectrum in redshift space, $P_0^s(k)$,

$$\frac{P_0^s(k)}{P^r(k)} = 1 + \frac{2}{3}f + \frac{1}{5}f^2, \tag{4.6}$$

where $P^r(k)$ denotes the matter power spectrum in real space. In practice, $P^r(k)$ cannot be obtained directly for a real survey without making approximations (e.g. Baugh and Efstathiou 1994).

In this chapter we also consider the estimator for f suggested by Cole et al. (1994), which is the ratio of quadrupole to monopole moments of the redshift space power spectrum, $P_2^s(k)/P_0^s(k)$. From Eq. 4.3 and after spherically averaging, the estimator for f is then

$$\frac{P_2^s(k)}{P_0^s(k)} = \frac{4f/3 + 4f^2/7}{1 + 2f/3 + f^2/5}, \tag{4.7}$$

which is independent of the real space power spectrum. Here, as before, $f = \beta/b$, with $b = 1$ for dark matter.

4.2.3 Modelling Non-Linear Distortions to the Power Spectrum in Redshift Space

Assuming the line of sight component is along the z-axis, the fully non-linear relation between the real and redshift space power spectrum can be written as Scoccimarro et al. (1999)

$$P^s(k, \mu) = \int \frac{\mathrm{d}^3 \mathbf{r}}{(2\pi)^3} e^{-i\mathbf{k}\cdot\mathbf{r}} \langle e^{i\lambda \Delta u_z} [\delta(\mathbf{x}) - f\nabla_z \cdot u_z(\mathbf{x})]$$

$$\times [\delta(\mathbf{x}') - f\nabla_z' \cdot u_z(\mathbf{x}')] \rangle, \tag{4.8}$$

where $\lambda = fk\mu$, u_z is the comoving peculiar velocity along the line of sight, $\Delta u_z = u_z(\mathbf{x}) - u_z(\mathbf{x}')$, $\mathbf{r} = \mathbf{x} - \mathbf{x}'$ and the only approximation made is the plane parallel approximation. This expression is the Fourier analog of the 'streaming model' first suggested by Peebles (1980) and modified by Fisher (1995) to take into account the density-velocity coupling. At small scales (as k increases) the exponential component damps the power, representing the impact of randomised velocities inside gravitationally bound structures.

Simplified models for redshift space distortions are frequently used. Examples include multiplying Eq. 4.6 by a factor which attempts to take into account small scale effects and is either a Gaussian or an exponential (Peacock and Dodds 1994). Two popular phenomenological examples of this which incorporates the damping effect of velocity dispersion on small scales is firstly the so called 'dispersion model' (Peacock and Dodds 1994),

$$P^s(k, \mu) = P^r(k)(1 + \beta\mu^2)^2 \frac{1}{(1 + k^2\mu^2\sigma_p^2/2)}, \tag{4.9}$$

and secondly the so called 'Gaussian model' (Peacock and Dodds 1994),

$$P^s(k, \mu) = P^r(k)(1 + \beta\mu^2)^2 \exp(-k^2\mu^2\sigma_p^2), \tag{4.10}$$

where σ_p is the pairwise velocity dispersion along the line of sight, which is treated as a parameter to be fitted to the data. Using numerical simulations, Hatton and Cole (1999) found a fit to the quadrupole to monopole ratio $P_2^s/P_0^s = (P_2^s/P_0^s)_{\text{lin}}(1-x^{1.22})$ to mimic damping and non-linear effects, where $(P_2^s/P_0^s)_{\text{lin}}$ is the linear theory prediction given by Eq. 4.7, $x = k/k_1$ and k_1 is a free parameter. They extended the dynamic range of simulations, to replicate the effect of a larger box, using the approximate method for adding long wavelength power suggested by Cole (1997).

The velocity divergence auto power spectrum is the ensemble average, $P_{\theta\theta} = \langle|\theta|^2\rangle$ where $\theta = \vec{\nabla}\cdot\vec{u}$ is the velocity divergence. The cross power spectrum of the velocity divergence and matter density is $P_{\delta\theta} = \langle|\delta\theta|\rangle$, where in this notation the matter density auto spectrum is $P_{\delta\delta} = \langle|\delta|^2\rangle$. In Eq. 4.8, the term in square brackets can be re-written in terms of these non-linear velocity divergence power spectra by multiplying out the brackets and using the fact that $\mu_i = \vec{k}_i\cdot\hat{z}/k_i$. Scoccimarro (2004) proposed the following model for the redshift space power spectrum in terms of $P_{\delta\delta}$, the non-linear matter power spectrum, $P_{\theta\theta}$ and $P_{\delta\theta}$,

$$P^s(k, \mu) = \left(P_{\delta\delta}(k) + 2f\mu^2 P_{\delta\theta}(k) + f^2\mu^4 P_{\theta\theta}(k)\right) \times e^{-(fk\mu\sigma_v)^2}, \tag{4.11}$$

where σ_v is the 1D linear velocity dispersion given by

$$\sigma_v^2 = \frac{1}{3} \int \frac{P_{\theta\theta}(k)}{k^2} d^3k. \tag{4.12}$$

In linear theory, $P_{\theta\theta}$ and $P_{\delta\theta}$ take the same form as $P_{\delta\delta}$ and depart from this at different scales. Using a simulation with 512^3 particles in a box of length $479\,h^{-1}$ Mpc

(Yoshida et al. 2001), Scoccimarro (2004) showed that this simple ansatz for $P_s(k, \mu)$ was an improvement over the Kaiser formula when comparing to N-body simulations in a ΛCDM cosmology. As this is a much smaller simulation volume than the one we use to investigate redshift space distortions we are able to test the fit to the measured power spectrum on much larger scales and to higher accuracy.

4.3 Results I: The Matter Power Spectrum in Real and Redshift Space

In Sects. 4.3.1 and 4.3.2 we present the redshift space distortions measured from the simulations in ΛCDM and quintessence cosmologies presented in Chap. 3, and we compare with the predictions of the linear and non-linear models discussed in Sects. 4.2.2 and 4.2.3.

4.3.1 Testing the Linear Theory Redshift Space Distortion

In the left panel of Fig. 4.2, we plot the ratio of the redshift space to real space power spectra, measured from the ΛCDM simulation at $z = 0$ and $z = 1$. Using the plane parallel approximation, we assume the observer is at infinity and as a result the velocity distortions are imposed along one direction in k-space. If we choose the line of sight direction to be the z-axis, for example, then $\mu = k_z/k$ where $k = |\vec{k}|$. In this chapter the power spectrum in redshift space represents the average of $P(k, \mu = k_x/k)$, $P(k, \mu = k_y/k)$ and $P(k, \mu = k_z/k)$ where the line of sight components are parallel to the x, y and z directions respectively. We use this average as there is a significant scatter in the amplitudes of the three redshift space power spectra on large scales, even for a computational box as large as the one we have used. The three monopoles of the redshift space power spectra $P(k, \mu = k_x/k)$, $P(k, \mu = k_y/k)$ and $P(k, \mu = k_z/k)$ measured in one of the realisations are plotted as the cyan, purple and red dashed lines respectively, to illustrate the scatter.

In Fig. 4.2 the Kaiser formula, given by Eq. 4.6, is plotted as a blue dotted line, using a value of $f = \Omega_m^{0.55}(z)$ for ΛCDM. The error bars plotted represent the scatter over four realisations after averaging over $P(k)$ obtained by treating the x, y and z directions as the line of sight. It is clear from this plot that the linear perturbation theory limit is only attained on extremely large scales ($k < 0.03\,\mathrm{hMpc}^{-1}$) at $z = 0$ and at $z = 1$. Non-linear effects are significant on scales $0.03 < k\,(\mathrm{hMpc}^{-1}) < 0.1$ which are usually considered to be in the linear regime. The measured variance in the matter power spectrum on these scales is $10^{-3} < \sigma^2 < 10^{-2}$.

In the right panel of Fig. 4.2 we plot the ratio P_2^s/P_0^s for ΛCDM at $z = 0$ and $z = 1$. The ratio agrees with the Kaiser limit (given in Eq. 4.6) down to smaller scales, $k < 0.06\,\mathrm{hMpc}^{-1}$, compared to the monopole ratio plotted in the left panel.

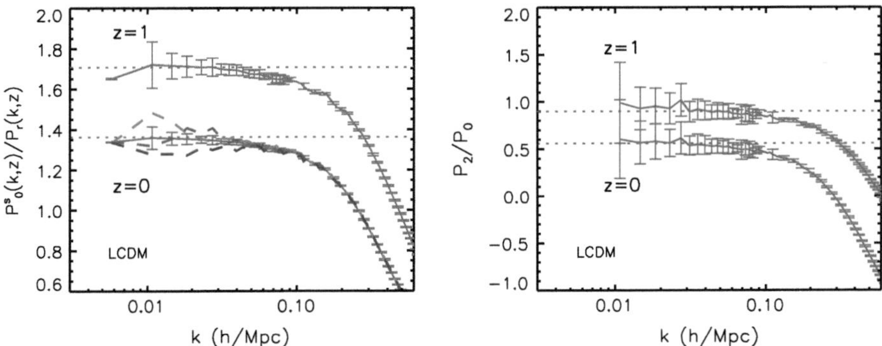

Fig. 4.2 *Left panel* the ratio of the monopole redshift power spectra and real space power spectra measured from the ΛCDM simulation at $z = 0$ and $z = 1$ are plotted as *blue lines*. The *error bars* plotted represent the scatter between the different power spectra from four ΛCDM simulations set up with different realisations of the density field with the distortions imposed along either the x, y or z axis and averaged. The power spectra $P(k, \mu = k_x/k)$, $P(k, \mu = k_y/k)$ and $P(k, \mu = k_z/k)$ measured from one simulation are plotted as the *cyan*, *purple* and *red dashed lines* respectively. *Right panel* the ratio of the quadrupole to monopole moment of the redshift space power spectrum measured from the simulations at $z = 0$ and $z = 1$ in ΛCDM are plotted in *blue*. It was not possible to accurately measure the quadrupole to monopole power in the first bin, so this point has not been plotted in the *right hand panel*. Note for wavenumbers $k > 0.1\,\mathrm{hMpc}^{-1}$, only every fifth *error bar* is plotted for clarity. The Kaiser formula, given by Eq. 4.6, is plotted as a *blue dotted line*. The *error bars* were obtained as described for the *left-hand panel*

Our results agree with previous work on the quadrupole and monopole moments of the redshift space power spectrum for ΛCDM (Cole et al. 1994; Hatton and Cole 1999; Scoccimarro 2004). At $z = 1$, the damping effects are less prominent and the Kaiser limit is attained over a slightly wider range of scales, $k < 0.1\,\mathrm{hMpc}^{-1}$, as non-linear effects are smaller then at $z = 0$. In the next section, we consider these ratios for the quintessence dark energy models in more detail. For each model we find that the analytic expression for the quadrupole to monopole ratio describes the simulation results over a wider range of wavenumber then the analogous result for the monopole moment.

4.3.2 Nonlinear Models of $P_s(k, \mu)$

The linear theory relationship between the real and redshift space power spectra given in Eq. 4.6 assumes various non-linear effects are small and can be neglected on large scales. These assumptions are listed in Sect. 4.2.2. In this section we consider the non-linear terms in the gradient of the line of sight velocity field and explore the scales at which it is correct to ignore such effects in the redshift space power spectrum. As a first step, we compare the model in Eq. 4.11, to measurements from N-body simulations for different quintessence dark energy models, without the damping term due to velocity dispersion. This will highlight the scale at which non-linear velocity divergence terms affect the matter power spectrum in redshift space and cause it to depart from the linear theory prediction.

If we rewrite $d\delta/d\tau$ as $aH(a)f(\Omega_m(a), \gamma)\,\delta$, where δ is the matter perturbation and τ is the conformal time, $dt = a(\tau)d\tau$, then the linear continuity equation becomes

$$\theta = \vec{\nabla}\cdot\vec{u} = -aHf\delta. \tag{4.13}$$

Throughout this chapter we normalise the velocity divergence as $\theta(k, a)/[-aH(a)$ $f(\Omega_m(a), \gamma)]$, so $\theta = \delta$ in the linear regime. The volume weighted velocity divergence power spectrum is calculated from the simulations according to the prescription given in Scoccimarro (2004). We interpolate the velocities and the densities onto a grid of 350^3 points and then measure the ratio of the interpolated momentum to the interpolated density field. In this way, we avoid having to correct for the CIC assignment scheme. A larger grid dimension could result in empty cells where $\delta \rightarrow 0$. A FFT grid of 350^3 was used to ensure all grid points had non-zero density and hence a well defined velocity at each point. We only plot the velocity power spectra in each of the figures up to half the Nyquist frequency for our default choice of $N_{\text{FFT}} = 350^3$, $k_{nq}/2 = \pi N_{\text{FFT}}/(2L_{\text{box}}) = 0.37\,\text{hMpc}^{-1}$ which is beyond the range typically used in BAO fitting when assuming linear theory.

The left panel in Fig. 4.3 shows the ratio of the power spectra, $P_{\delta\delta}$, $P_{\delta\theta}$ and $P_{\theta\theta}$ measured at $z = 0$, to the power spectra measured at $z = 5$ scaled using the ratio of the square of the linear growth factor at $z = 5$ and $z = 0$ for ΛCDM. It is clear from this plot that all $P(k)$ evolve as expected in linear theory on the largest scales. Note a linear scale is used on the x-axis in this case. In the right panel in Fig. 4.3 all the power spectra have been divided by the linear theory matter power spectrum measured from the simulation at $z = 5$, scaled using the ratio of the linear growth factor at $z = 5$ and $z = 0$. This removes the sampling variance from the plotted ratio (Baugh and Efstathiou 1994). In both panels, the error bars represent the scatter over eight simulations in ΛCDM averaging the power spectra after imposing the distortions along the x, y or z axis in turn. From this figure we can see that the non-linear velocity divergence power spectra can be substantially different from the matter power spectrum on very large scales $k \sim 0.03\,\text{hMpc}^{-1}$. The linear perturbation theory assumption that the velocity divergence power spectra is the same as the matter $P(k)$ is not valid even on these large scales. In the case of ΛCDM this difference is $\sim$20 % at $k = 0.1\,\text{hMpc}^{-1}$. Note in the right panel in Fig. 4.3, the 10 % difference in the ratio of the cross power spectrum to the matter power spectrum, on the largest scale considered, indicates that we have a biased estimator of θ which is low by approximately 10 %.

We find that the $P_{\delta\theta}$ and $P_{\theta\theta}$ measured directly from the simulation differ from the matter power spectrum by more then was reported by Percival and White (2009). These authors did not measure $P_{\delta\theta}$ and $P_{\theta\theta}$ directly, but instead obtained these quantities by fitting Eq. 4.14 to the redshift space monopole power spectrum measured from the simulations. In Fig. 4.4 we plot the same ratios as shown in the right panel of Fig. 4.3 measured from one ΛCDM (left panel) and SUGRA (right panel) simulation.

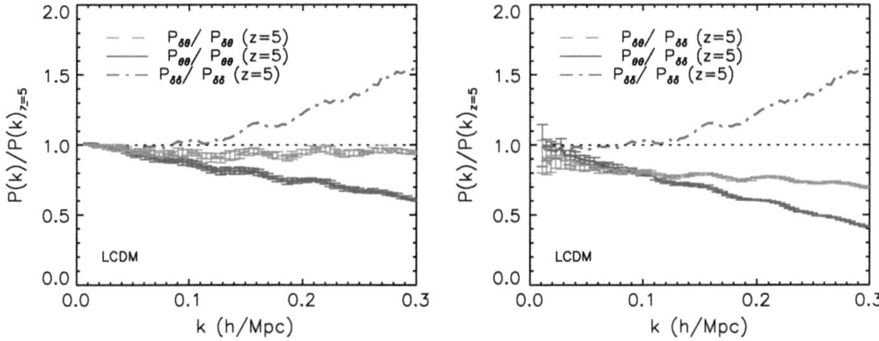

Fig. 4.3 *Left panel* the ratio of the non-linear power spectra, $P_{\delta\delta}$, $P_{\delta\theta}$ and $P_{\theta\theta}$ for ΛCDM measured from the simulation at $z = 0$, divided by the corresponding power spectrum measured from the simulation at $z = 5$, scaled using the square of the ratio of the linear growth factor at $z = 5$ and $z = 0$. The non-linear matter power spectrum is plotted as a *grey dot-dashed line*, the non-linear velocity divergence auto power spectrum $P_{\theta\theta}$ is plotted as a *blue solid line* and the non-linear cross power spectrum, $P_{\delta\theta}$, is plotted as a *green dashed line*. *Right panel* the ratio of the non-linear power spectra, $P_{\delta\delta}$, $P_{\delta\theta}$ and $P_{\theta\theta}$, to the linear theory matter $P(k)$ in ΛCDM measured from the simulation at $z = 0$. All power spectra have been divided by the linear theory matter power spectrum measured from the simulation at $z = 5$, scaled using the square of the ratio of the linear growth factor at $z = 5$ and $z = 0$. In both panels the *error bars* represent the scatter over eight ΛCDM realisations after imposing the peculiar velocity distortion along each Cartesian axis in turn

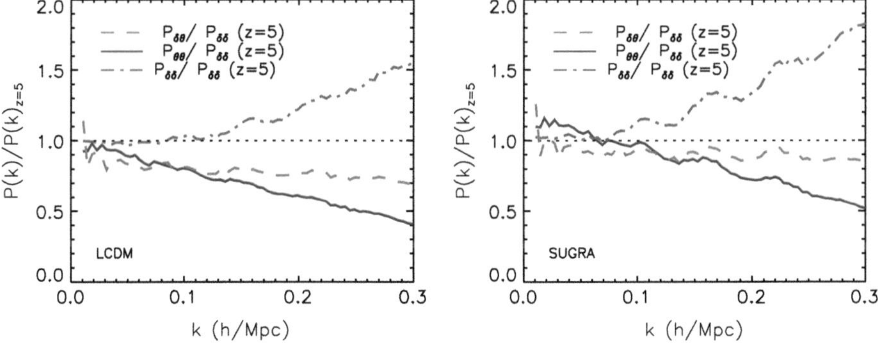

Fig. 4.4 *Left panel* the ratio of the non-linear power spectra, $P_{\delta\delta}$, $P_{\delta\theta}$ and $P_{\theta\theta}$, to the linear theory $P(k)$ in ΛCDM measured from one realisation of the matter density and velocity fields at $z = 0$. All power spectra have been divided by the linear theory matter power spectrum measured from the simulation at $z = 5$, scaled using the square of the ratio of the linear growth factor at $z = 5$ and $z = 0$. *Right panel* similar to that in the *left panel* but for the SUGRA quintessence model. The lines are the same as used in the *left hand panel*

From our simulations it is possible to find a realisation of the density and velocity fields where the measured matter power spectrum and the velocity divergence power spectra are similar on large scales.

Having found that the measured $P_{\delta\theta}$ and $P_{\theta\theta}$ differ significantly from $P_{\delta\delta}$, we now test if the grid assignment scheme has any impact on our results. As explained

Fig. 4.5 A comparison of the impact of the FFT grid dimension on power spectrum estimation. The plots show the ratio of the non-linear power spectra, $P_{\theta\theta}$ (*upper panel*) and $P_{\delta\theta}$ (*lower panel*), to the linear theory matter power spectrum measured from the simulations in ΛCDM, using different FFT grid sizes. From *bottom* to *top* in each panel the lines show the ratios for grid sizes $N_{\mathrm{FFT}} = 128$ (*purple*), $N_{\mathrm{FFT}} = 256$ (*blue*), $N_{\mathrm{FFT}} = 350$ (*red*) and $N_{\mathrm{FFT}} = 375$ (*green*)

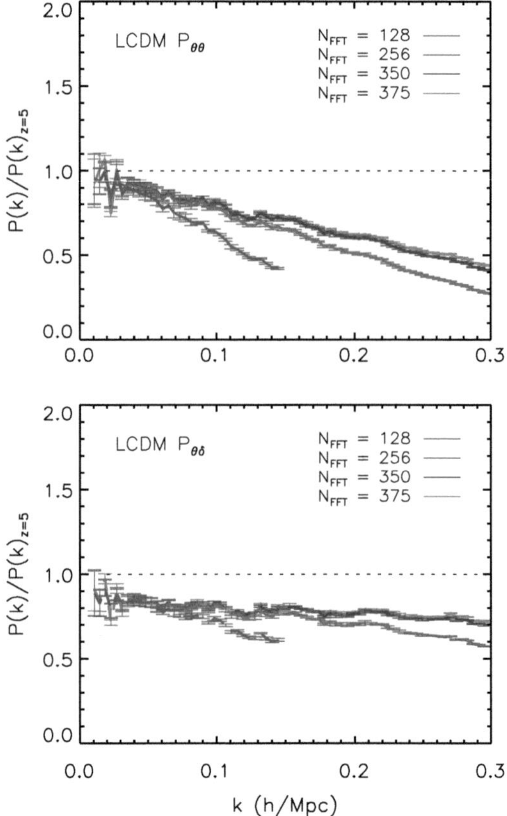

in Chap. 2, the velocity $P(k)$ are computed by taking the Fourier transform of the momentum field divided by the density field to reduce the impact of the grid assignment scheme (Scoccimarro 2004). Pueblas and Scoccimarro (2009) showed that the CIC assignment scheme affects the measured $P(k)$ beyond $\sim 20\%$ of the Nyquist frequency. In Fig. 4.5 we show the power spectrum measurements for four different FFT dimensions to show the scales at which we get a robust measurement. For $N_{\mathrm{FFT}} = 350$ the power spectra have converged on scales up to $k \sim 0.2\,\mathrm{hMpc}^{-1}$.

In the top row of Fig. 4.6, the ratios $P_0^s(k)/P^r(k)$ and $P_2^s(k)/P_0^s(k)$ are plotted as solid lines in the left and right hand panels respectively. In this figure we have overplotted as grey dashed lines, the ratio of the redshift space monopole moment to the real space power spectrum where

$$P_0^s(k) = P_{\delta\delta}(k) + \frac{2}{3} f P_{\delta\theta}(k) + \frac{1}{5} f^2 P_{\theta\theta}(k). \tag{4.14}$$

On scales $0.05 < k\,(\mathrm{hMpc}^{-1}) < 0.2$, this model for the redshift space power spectrum reproduces the measured $P_s(k, \mu)$ and is a significant improvement compared

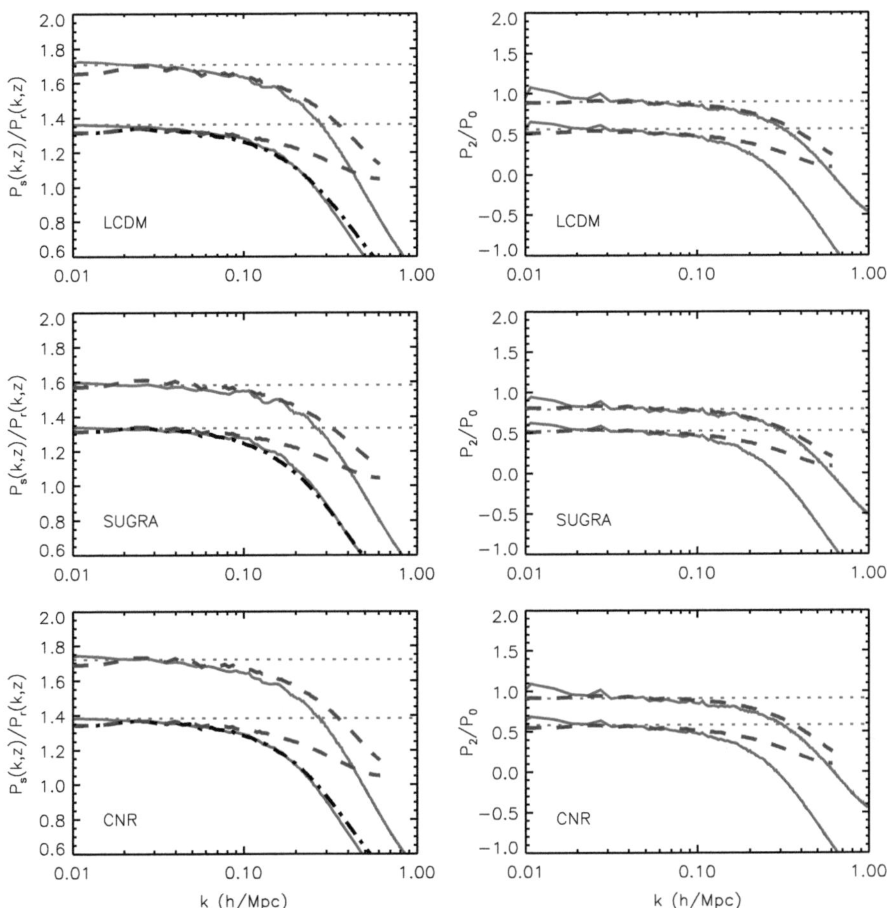

Fig. 4.6 The *left hand column* shows the ratio of the monopole of redshift power spectra to the real space power spectra at $z = 0$ and $z = 1$. The *right hand column* shows the ratio of the quadrupole to monopole moment of the redshift space power spectra at $z = 0$ and $z = 1$. Different rows show different dark energy models as labelled. *Top row* the ratio of the redshift and real space power spectra in ΛCDM are plotted as *solid lines* in the *left panel*. The *dashed lines* represent the same ratio using Eq. 4.14 for the monopole of the redshift space power spectrum. The *dot-dash line* represents the model given in Eq. 4.11 which includes velocity dispersion effects. In the *right panel* the ratio of the quadrupole to monopole moment of the redshift space power spectra in ΛCDM are plotted as *solid lines*. The same ratio using Eq. 4.15 for the redshift space power spectrum is plotted as *dashed lines*. *Middle row* same as the *top row* but for the SUGRA quintessence model. *Bottom row* same as the *middle row* but for the CNR quintessence model

to Eq. 4.6. This form does not include any modelling of the damping due to velocity dispersion. The extended model proposed by Scoccimarro (2004) given in Eq. 4.11, which does include damping, is also plotted as a black dot-dashed line for ΛCDM in the top row in Fig. 4.6. The redshift space quadrupole to monopole ratio in the quasi-linear regime, including the velocity divergence power spectra, is

$$\frac{P_2^s}{P_0^s} = \frac{\frac{4}{3} f P_{\delta\theta} + \frac{4}{7} f^2 P_{\theta\theta}}{P_{\delta\delta} + \frac{2}{3} f P_{\delta\theta} + \frac{1}{5} f^2 P_{\theta\theta}}. \tag{4.15}$$

This model does well at reproducing the ratio of the redshift space to real space power spectrum, although it underpredicts the ratio on scales $k < 0.02\,\mathrm{hMpc}^{-1}$. The corresponding plots for the SUGRA and CNR models are shown in the middle and bottom rows of Fig. 4.6. It is clear that including the velocity divergence power spectrum in the model for P_0^s and P_2^s, produces a good fit to the measured redshift space power in both quintessence models on scales up to $k \sim 0.2\,\mathrm{hMpc}^{-1}$.

4.4 Results II: The Density Velocity Relation

In Sect. 4.4.1 we examine the relationship between the non-linear matter and velocity divergence power spectra in different cosmologies. In Sect. 4.4.2 we study the redshift dependence of this relationship and provide a prescription which can be followed to generate predictions for the non-linear velocity divergence power spectrum at a given redshift.

4.4.1 Dependence on Cosmological Model

The linear continuity equation, Eq. 4.13, gives a one to one correspondence between the velocity and density fields with a cosmology dependent factor, $f(\Omega_m, \gamma)$. Once the overdensities become non-linear, this relationship no longer holds. Bernardeau (1992) derived the non-linear relation between δ and θ in the case of an initially Gaussian field. Chodorowski and Lokas (1997) extended this relation into the weakly non-linear regime up to third order in perturbation theory and found the result to be a third order polynomial in θ. More recently, Bilicki and Chodorowski (2008) found a relation between θ and δ using the spherical collapse model. In all of these relations, the dependence on cosmological parameters was found to be extremely weak (Bernardeau 1992; Bouchet et al. 1995). The velocity divergence depends on Ω_m and Ω_Λ, in a standard ΛCDM cosmology, only through the linear growth rate, f (Scoccimarro et al. 1999).

We showed in the previous section that including the velocity divergence auto and cross power spectrum accurately reproduces the redshift space power spectrum for a range of dark energy models on scales where the Kaiser formula fails. The quantities in Eqs. 4.15 and 4.11 can be calculated if we exploit the relationship between the velocity and density field. In Fig. 4.7 we plot the velocity divergence auto (left panel) and cross (right panel) power spectrum as a function of the matter power spectrum for ΛCDM and the three quintessence dark energy models. We find that the density velocity relationship is very similar for each model at the redshifts considered, with only a slight difference for the SUGRA model at high redshifts and at small scales. The departure of the SUGRA model from the general density velocity relation is due to shot noise, which affects the power spectrum most at these scales in the SUGRA model as it has the lowest amplitude. We have verified that this effect is due to shot noise by sampling half the particles in the same volume, thereby doubling the shot noise, and repeating the $P(k)$ measurement to find an even larger departure. Fig. 4.7 shows the independence of the density velocity relation not only of the values of cosmological parameters, as found in previous works, Bernardeau (1992), but also a lack of dependence on the cosmological expansion history and initial power spectrum.

Fitting over the range $0.01 < k\,(\text{h/Mpc}) < 0.3$, we find the following function accurately describes the relation between the non-linear velocity divergence and matter power spectrum at $z = 0$ to better than 5 % on scales $k < 0.3\,\text{hMpc}^{-1}$,

$$P_{xy}(k) = g(P_{\delta\delta}(k)) = \frac{\alpha_0\sqrt{P_{\delta\delta}(k)} + \alpha_1 P_{\delta\delta}^2(k)}{\alpha_2 + \alpha_3 P_{\delta\delta}(k)}, \tag{4.16}$$

where $P_{\delta\delta}$ is the non-linear matter power spectrum. For the cross power spectrum $P_{xy} = P_{\delta\theta}$, $\alpha_0 = -12288.7$, $\alpha_1 = 1.43$, $\alpha_2 = 1367.7$ and $\alpha_3 = 1.54$ and for $P_{xy} = P_{\theta\theta}$, $\alpha_0 = -12462.1$, $\alpha_1 = 0.839$, $\alpha_2 = 1446.6$ and $\alpha_3 = 0.806$; all points were weighted equally in the fit and the units for α_0, α_1 and α_3 are $(\text{Mpc/h})^{3/2}$, $(\text{Mpc/h})^{-3}$ and $(\text{Mpc/h})^{-3}$ respectively. The power spectra used for this fit are the average $P_{\theta\theta}$, $P_{\delta\theta}$ and $P_{\delta\delta}$ measured from eight ΛCDM simulations.

4.4.2 Approximate Formulae for $P_{\delta\theta}$ and $P_{\theta\theta}$ for Arbitrary Redshift

In perturbation theory, the solution for the density contrast is expanded as a series around the background value. Scoccimarro et al. (1998) found the following solutions for δ and θ to arbitrary order in perturbation theory,

$$\delta(k, \tau) = \sum_{n=1}^{\infty} D_n(\tau)\delta_n(k)$$

$$\theta(k, \tau) = \sum_{n=1}^{\infty} E_n(\tau)\theta_n(k), \tag{4.17}$$

where $\delta_1(k)$ and $\theta_1(k)$ are linear in the initial density field, δ_2 and θ_2 are quadratic in the initial density field etc. Scoccimarro et al. (1998) showed that using a simple approximation to the equations of motion, $f(\Omega_m) = \Omega_m^{1/2}$, the equations become

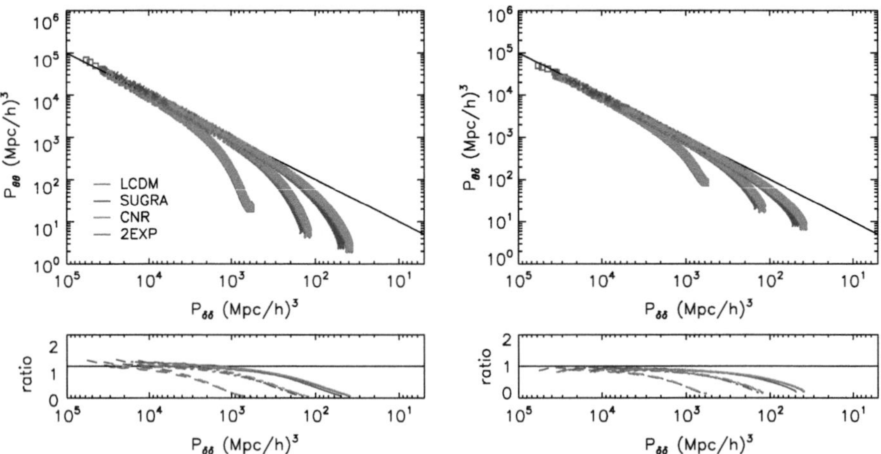

Fig. 4.7 Non-linear velocity divergence auto (*left*) and cross (*right*) power spectrum plotted as a function of the non-linear matter power spectrum at $z = 0$, 1 and 2 in three quintessence models and ΛCDM, as labelled. The ratio of the velocity divergence power spectra to the matter power spectrum at each redshift is plotted in the *smaller panels* beneath each *main panel*

separable and $E_n(\tau) = D_n(\tau) = D(\tau)^n$, where $D(\tau)$ is the linear growth factor of density perturbations. We shall use these solutions for $\delta(k, \tau)$ and $\theta(k, \tau)$ to approximate the redshift dependence of the density velocity relation found in Sect. 4.4.1. This relation does not depend on the cosmological model but we shall assume a ΛCDM cosmology and find the approximate redshift dependence as a function of the ΛCDM linear growth factor.

The fitting function given in Eq. 4.16 generates the non-linear velocity divergence power spectrum, $P_{\delta\theta}$ or $P_{\theta\theta}$ from the non-linear matter power spectrum, $P_{\delta\delta}$ at $z = 0$. Figure 4.8 shows a simple illustration of how the function $g(P_{\delta\delta})$ and $P_{\delta\delta}$ at $z = 0$ can be rescaled to give the velocity divergence power spectra at a higher redshift, z'. Using the simplified notation in the diagram, where $P_1 = P_{\delta\delta}$, and given the function $g(P_{\delta\delta})$, we can find a redshift dependent function, $c(z)$, with which to rescale $g(P_{\delta\delta}(z = 0))$ to the velocity divergence $P(k)$ at z'. At the higher redshift, z', the non-linear matter and velocity divergence power spectra are denoted as P'_1 and P'_2 respectively in Fig. 4.8.

Using the solutions in Eq. 4.17, to third order in perturbation theory, see Appendix B.1, we assume a simple expansion with respect to the initial density field, to find the following ansatz for the mapping $P'_1(z = z') \rightarrow P'_2(z = z')$ which can be approximated as $P_1(z = 0)/c^2(z = 0, z') \rightarrow g(P_1)/c^2(z = 0, z')$ where

$$c(z, z') = \frac{D(z) + D^2(z) + D^3(z)}{D(z') + D^2(z') + D^3(z')}, \tag{4.18}$$

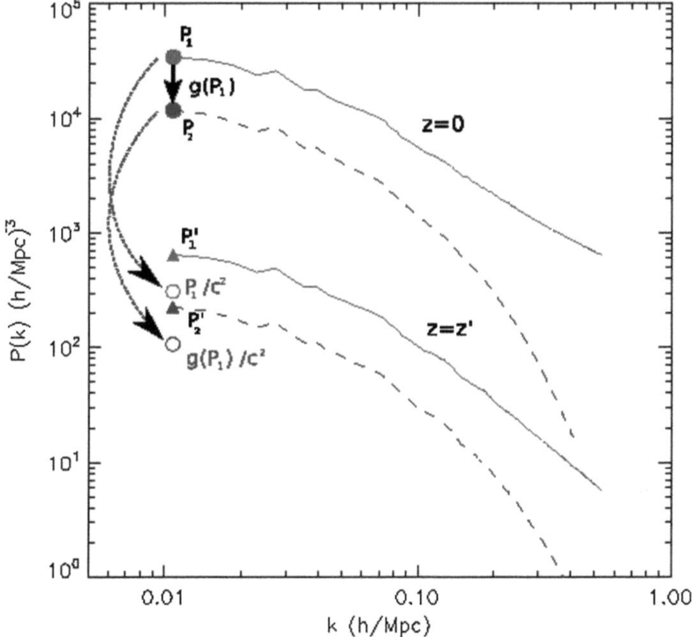

Fig. 4.8 A schematic illustration showing how the $z = 0$ non-linear matter power spectrum can be rescaled to find the velocity divergence power spectrum at any redshift $z = z'$. The *upper two curves* represent the non-linear matter power spectrum, P_1, in *grey* and the velocity divergence power spectrum, P_2, plotted as a *blue dashed line*, at $z = 0$. The power in the first bin is represented as a *filled circle* for each spectrum. The *lower two curves*, P_1' and P_2', are the non-linear matter and velocity divergence spectra at $z = z'$. The power in the first bin is represented as a *filled triangle* in each case. The fitting formula for $g(P_1)$ (Eq. 4.16) generates the non-linear velocity divergence power spectra at $z = 0$. Using the function given in Eq. 4.18, the matter power spectrum P_1 and $g(P_1)$ can be rescaled to an earlier redshift. The power in the first bin from the rescaled P_1 and $g(P_1)$ are shown as an *empty grey* and *blue circle* respectively. Note that P_1 and P_2 have been artificially separated for clarity

and $D(z)$ is the linear growth factor. The equivalence of these mappings gives $P_1' - P_2' = (P_1 - g(P_1))/c^2$ which allows us to calculate P_2' at $z = z'$ if we have $P_1(z = 0)$, $g(P_1(z = 0)$ and $P_1'(z = z')$. Writing this now in terms of $P_{\delta\delta}$, instead of P_1, we have the following equation

$$P_{xy}(k, z') = \frac{g(P_{\delta\delta}(k, z = 0)) - P_{\delta\delta}(k, z = 0)}{c^2(z = 0, z')} + P_{\delta\delta}(k, z'), \qquad (4.19)$$

where $g(P_{\delta\delta})$ is the function in Eq. 4.16 and P_{xy} is either the nonlinear cross or auto power spectrum, $P_{\delta\theta}$ or $P_{\delta\delta}$.

In the left panel of Fig. 4.9, we plot the ΛCDM non-linear power spectrum $P_{\theta\theta}$ at $z = 0, 1, 2$ and 3. The function given in Eq. 4.19 is also plotted as red dashed lines

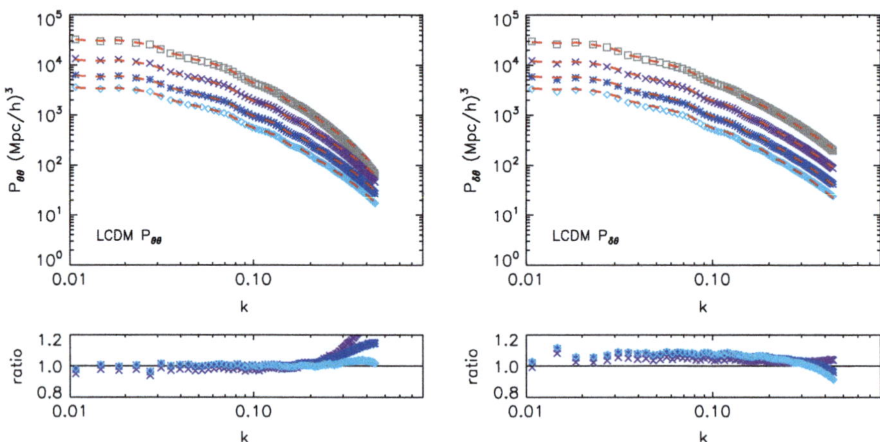

Fig. 4.9 Non-linear velocity divergence auto and cross power spectrum, in the *left* and *right panels* respectively, measured from the ΛCDM simulations at $z = 0$ (*open grey squares*), $z = 1$ (*purple crosses*), $z = 2$ (*blue stars*) and $z = 3$ (*cyan diamonds*). Overplotted as *red dashed lines* is the function given in Eq. 4.19 at redshifts $z = 1$, 2 and 3. The *lower panels* show the function in Eq. 4.19 divided by the measured spectra at $z = 1$, 2 and 3

using the factor $c(z, z')$ given in Eq. 4.18 and the ΛCDM linear growth factor at redshift $z = 0, 1, 2$ and 3 respectively. The ratio plot shows the difference between the exact $P_{\theta\theta}$ power spectrum and the function given in Eq. 4.19. The right panel in Fig. 4.9 shows a similar plot for the $P_{\delta\theta}$ power spectrum. In both cases we find very good agreement between the scaled fitting formula and the measured power spectrum. Scaling the $z = 0$ power spectra using this approximation in Eq. 4.18 reproduces the non-linear $z = 1, 2$ and 3, $P_{\delta\theta}$ to $\sim 5\,\%$ and $P_{\theta\theta}$ to better than $5\,\%$ on scales $0.05 < k(\mathrm{hMpc}^{-1}) < 0.2$. It is remarkable that scaling the $z = 0$ fitting formula using c in Eq. 4.18 works so well at the different redshifts up to $k < 0.3\,\mathrm{h/Mpc}$ and is completely independent of scale.

To summarise the results of this section we have found that the quadrupole to monopole ratio given in Eq. 4.15 and the model in Eq. 4.11, which includes the non-linear matter and velocity divergence power spectra at a given redshift z', can be simplified by using the following prescription. Assuming a cosmology with a given linear theory matter power spectrum we can compute the non-linear matter $P(k)$ at $z = 0$ and at the required redshift, z', using, for example, the phenomenological model HALOFIT (Smith et al. 2003) or the method proposed by Casarini et al. (2009) in the case of quintessence dark energy. These power spectra can then be used in Eq. 4.19 together with the function g, given in Eq. 4.16, and the linear theory growth factor between redshift $z = 0$ and $z = z'$ to find the velocity divergence auto or cross power spectrum. As can be seen from Fig. 4.9 the function given in Eq. 4.19 agrees with the measured non-linear velocity divergence power spectrum to $\sim 10\,\%$ for $k < 0.3\mathrm{hMpc}^{-1}$ and to $< 5\,\%$ for $k < 0.2\mathrm{hMpc}^{-1}$ for ΛCDM. We have verified that this prescription also reproduces $P_{\delta\theta}$ and $P_{\theta\theta}$ to an accuracy of $10\,\%$

for $k < 0.3 \mathrm{hMpc}^{-1}$ for the CNR, SUGRA and 2EXP models using the corresponding matter power spectrum and linear growth factor for each model. This procedure simplifies the redshift space power spectrum in Eq. 4.11 and the quadrupole to monopole ratio given in Eq. 4.15. For the dark energy models considered in this chapter, this ratio provides an improved fit to the redshift space $P(k, \mu)$ compared to the Kaiser formula and incorporating the density velocity relation eliminates any new parameters which need to be measured separately and may depend on the cosmological model.

4.5 Summary

We use simulations of three quintessence dark energy models which have different expansion histories, linear growth rates and power spectra compared to ΛCDM. In Chap. 3, Jennings et al. (2010), we carried out the first fully consistent N-body simulations of quintessence dark energy, taking into account different expansion histories, linear theory power spectra and best fitting cosmological parameters Ω_m, Ω_b and H_0, for each model. In this chapter we examine the redshift space distortions in the SUGRA, CNR and 2EXP quintessence models. These models are representative of a broader class of quintessence models which have different growth histories and dark energy densities at early times compared to ΛCDM. In particular the SUGRA model has a linear growth rate that differs from ΛCDM by $\sim$20 % at $z = 5$ and the CNR model has high levels of dark energy at early times, $\Omega_\mathrm{DE} \sim 0.03$ at $z \sim 200$. The 2EXP model has a similar expansion history to ΛCDM at low redshifts, $z < 5$, despite having a dynamical equation of state for the dark energy component.

Redshift space distortions observed in galaxy surveys are the result of peculiar velocities which are coherent on large scales, leading to a boost in the observed redshift space power spectrum compared to the real space power spectrum (Kaiser 1987). On small scales these peculiar velocities are incoherent and give rise to a damping in the ratio of the redshift to real space power spectrum. The Kaiser formula is a prediction of the boost in this ratio on very large scales, where the growth is assumed to be linear, and can be expressed as a function of the linear growth rate and bias, neglecting all non-linear contributions.

In previous work, using N-body simulations in a periodic cube of $300\,\mathrm{h}^{-1}\mathrm{Mpc}$ on a side, Cole et al. (1994) found that the measured value of $\beta = f/b$, where b is the linear bias, deviates from the Kaiser formula on wavelengths of $50\,\mathrm{h}^{-1}$ Mpc or more as a result of these non-linearities. Hatton and Cole (1998) extended this analysis to slightly larger scales using the Zel'dovich approximation combined with a dispersion model where non-linear velocities are treated as random perturbations to the linear theory velocity. These previous studies do not provide an accurate description of the non-linearities in the velocity field for two reasons. Firstly, the Zel'dovich approximation does not model the velocities correctly, as it only treats part of the bulk motions. Secondly, in a computational box of length $300\,\mathrm{h}^{-1}\mathrm{Mpc}$, the power which determines the bulk flows has not converged. In this thesis we use a large computational box of side $1500\,h^{-1}\mathrm{Mpc}$, which allows us to measure redshift space distortions on large scales to far greater accuracy than in previous work.

In this chapter we find that the ratio of the monopole of the redshift space power spectrum to the real space power spectrum agrees with the linear theory Kaiser formula only on extremely large scales $k < 0.03\,\mathrm{hMpc}^{-1}$ in both ΛCDM and the quintessence dark energy models. We still find significant scatter between choosing different axes as the line of sight, even though we have used a much larger simulation box than that employed in previous studies. As a result we average over the three power spectra, assuming the distortions lie along the x, y and z directions in turn, for the redshift space power spectrum in this chapter. Instead of using the measured matter power spectrum in real space, we find that the estimator suggested by Cole et al. (1994), involving the ratio of the quadrupole to monopole redshift space power spectrum, works better than using the monopole and agrees with the expected linear theory on slightly smaller scales $k < 0.07\,\mathrm{hMpc}^{-1}$ at $z = 0$ for both ΛCDM and the quintessence models.

As the measured redshift space distortions only agree with the Kaiser formula on scales $k < 0.07\,\mathrm{hMpc}^{-1}$, it is clear that the linear approximation is not correct on scales which are normally considered to be in the 'linear regime', $k < 0.2\,\mathrm{hMpc}^{-1}$. In linear theory, the velocity divergence power spectrum is simply a product of the matter power spectrum and the square of the linear growth rate. In this thesis we have demonstrated that non-linear terms in the velocity divergence power spectrum persist on scales $0.04 < k\,(\mathrm{hMpc}^{-1}) < 0.2$. These results agree with Scoccimarro (2004) who also found significant non-linear corrections due to the evolution of the velocity fields on large scales, assuming a ΛCDM cosmology. We have shown that including the non-linear velocity divergence auto and cross power spectrum in the expression for the redshift space $P(k)$ leads to a significant improvement when trying to match the measured quadrupole to monopole ratio for both ΛCDM and quintessence dark energy models.

Including the non-linear velocity divergence cross and auto power spectra in the expression for the redshift space power spectrum increases the number of parameters needed and depends on the cosmological model that is used. Using the non-linear matter and velocity divergence power spectra we have found a density velocity relation which is model independent over a range of redshifts. Using this relation it is possible to write the non-linear velocity divergence auto or cross power spectrum at a given redshift, z', in terms of the non-linear matter power spectrum and linear growth factor at $z = 0$ and $z = z'$. This formula is given in Eq. 4.19 in Sect. 4.4.2. We find that this formula accurately reproduces the non-linear velocity divergence $P(k)$ to within 10 % for $k < 0.3\,\mathrm{hMpc}^{-1}$ and to better than 5 % for $k < 0.2\,\mathrm{hMpc}^{-1}$ for both ΛCDM and the dark energy models used in this chapter. It is clear that including the non-linear velocity divergence terms results in an improved model for redshift space distortions on scales $k < 0.2\,\mathrm{hMpc}^{-1}$ for different cosmological models.

References

Baugh CM, Efstathiou G (1994) MNRAS 270:183
Bernardeau F (1992) Astrophys J 390:L61
Bilicki M, Chodorowski MJ (2008) MNRAS 391:1796
Bouchet FR, Colombi S, Hivon E, Juszkiewicz R (1995) Astron Astrophys 296:575
Casarini L, Macciò AV, Bonometto SA (2009) JCAP 3:14
Chodorowski MJ, Lokas EL (1997) MNRAS 287:591
Cole S (1997) MNRAS 286:38
Cole S, Fisher KB, Weinberg DH (1994) MNRAS 267:785
Doran M, Robbers G (2006) JCAP 0606:026
Fisher KB (1995) Astrophys J 448:494
Guzzo L et al (2008) Nature 451:541
Hamilton AJS (1998) Astrophys Sp Sci 231:185
Hatton S, Cole S (1999) MNRAS 310:1137
Hatton SJ, Cole S (1998) MNRAS 296:10
Jackson J (1972) MNRAS 156:1P
Jennings E, Baugh CM, Angulo RE, Pascoli S (2010) MNRAS 401:2181
Kaiser N (1987) MNRAS 227:1
Lahav O, Lilje PB, Primack JR, Rees MJ (1991) MNRAS 251:128
Linder EV (2005) Phys Rev D 72:043529
Linder EV, Cahn RN (2007) Astropart Phys 28:481
Linder EV (2008) Astropart Phys 29:336
Linder EV (2009) Phys Rev 79:063519
Peacock JA, Dodds SJ (1994) MNRAS 267:1020
Peebles PJE (1976) Astrophys J 205:318
Peebles PJE (1980), The large scale structure of the universe. Princeton University Press, Princeton
Percival WJ, White M (2009) MNRAS 393:297
Pueblas S, Scoccimarro R (2009) Phys Rev D 80:043504
Scoccimarro R (2004) Phys Rev D 70:083007
Scoccimarro R, Couchman HMP, Frieman JA (1999) Astrophys J 517:531
Scoccimarro R et al (1998) Astrophys J 496:586
Simpson F, Peacock JA (2010) Phys Rev D 81:043512
Smith RE et al (2003) MNRAS 341:1311
Song Y, Percival WJ (2009) JCAP 10:4
Stril A, Cahn RN, Linder EV (2009) arXiv:0910.1833
Wang Y (2008) JCAP 0805:021
Wang L-M, Steinhardt PJ (1998) Astrophys J 508:483
White M, Song Y, Percival WJ (2009) MNRAS 397:1348
Yoshida N, Sheth RK, Diaferio A (2001) MNRAS 328:669

Chapter 5
Testing Gravity Using the Growth of Large Scale Structure in the Universe

5.1 Introduction

Dark energy and modified gravity models can produce similar expansion histories for the Universe, which can be derived from the Hubble parameter measured, for example, using Type Ia SN. The expansion history of the Universe in dark energy and modified gravity cosmologies can be described using an effective equation of state. If two models have the same equation of state, as a consequence, it is not possible to distinguish between them using measurements of the expansion history alone. However, cosmic structures are expected to collapse under gravity at different rates in the dark energy and modified gravity cosmologies.

The growth rate is a measure of how rapidly overdense regions are collapsing under gravity to form large structures in the Universe. Dark energy or modified gravity models predict different growth rates for the large scale structure of the Universe, which can be measured using redshift space distortions of clustering. As noted by Linder (2005), in the case of general relativity, the second order differential equation for the growth of density perturbations depends only on the expansion history through the Hubble parameter, $H(a)$, or the equation of state, $w(a)$. This is not the case for modified gravity theories. By comparing the cosmic expansion history with the growth of structure, it is possible to distinguish the physical origin of the accelerating expansion of the Universe as being due either to dark energy or modified gravity (Lue et al. 2004; Linder 2005). If there is no discrepancy between the observed growth rate and the theoretical prediction assuming general relativity, this implies that a dark energy component alone can explain the accelerated expansion. We test this assumption using large volume N-body simulations which are the only way to accurately follow the growth of cosmic structure and probe the limits of linear perturbation theory.

This chapter is organised as follows: In Sect. 5.2.1 we discuss the linear growth rate and its dependence on the cosmological model. In Sect. 5.2.2 we consider modified gravity models which feature a time varying Newton's constant. Our main results are presented in Sect. 5.3. The details of the simulations and tests of the code are

E. Jennings, *Simulations of Dark Energy Cosmologies*, Springer Theses, 83
DOI: 10.1007/978-3-642-29339-9_5, © Springer-Verlag Berlin Heidelberg 2012

given in Sect. 5.3.1. In Sect. 5.3.2 we present the measured redshift space distortions in the modified gravity and quintessence dark energy simulations. In Sect. 5.3.3 we test several models for the redshift space power spectrum and determine the best fit value of the growth rate for each model fitting over different intervals in Fourier space.

5.2 Testing Modifications to General Relativity

5.2.1 The Linear Growth Rate

The rate at which large scale structures grow is driven by two opposing mechanisms: gravitational instability (set by Newton's constant, G_N) and the expansion rate of the Universe (given by $H(a)$). In the framework of general relativity, the growth of a density fluctuation, $\delta \equiv (\rho(x,t) - \bar{\rho})/\bar{\rho}$, where $\bar{\rho}_m$ is the average matter density, depends only on the expansion history, $H(a)$. In alternative theories of gravity, e.g. where the modifications can be parametrized by a time-varying gravitational constant, $\tilde{G}$, the growth of perturbations will depend on *both* this varying gravitational coupling and the expansion history. By using the measured expansion history to predict the growth of structure and comparing this to a direct measurement of the growth rate, it may be possible to determine whether the physical origin of the accelerating cosmic expansion is due either to dark energy or modified gravity (Lue et al. 2004; Linder 2005).

Using the perturbed equations of motion, within general relativity, the growth of density perturbations evolves according to

$$\ddot{\delta} + 2H\dot{\delta} - 4\pi G_N \rho_m \delta = 0, \tag{5.1}$$

where the matter overdensity $\delta = \rho_m/\bar{\rho} - 1$, H is the Hubble expansion rate, G_N is the present value of the gravitational constant found in laboratory experiments and a dot denotes a derivative with respect to time. If we change variables to $g = \delta/a$ and allow the gravitational constant $\tilde{G}$ to vary in time, this equation becomes (Linder 2005)

$$\frac{d^2 g}{da^2} + \left(5 + \frac{1}{2}\frac{d\ln H^2}{d\ln a}\right)\frac{1}{a}\frac{dg}{da}$$
$$+ \left(3 + \frac{1}{2}\frac{d\ln H^2}{d\ln a} - \frac{3}{2}\frac{\tilde{G}(a)}{G_N}\Omega_m(a)\right)g = 0, \tag{5.2}$$

where $\Omega_m(a)$ is the ratio of the matter density to the critical density as a function of scale factor, a. It is clear from Eq. 5.2 that in the framework of general relativity, $\tilde{G}(a)/G_N = 1$ and the growth of perturbations depends only on the expansion history, $H(a)$. In theories of modified gravity the growth of perturbations will depend on both the expansion history and $\tilde{G}(a)$.

5.2.2 Time Variation of Newton's Constant

Modifications of general relativity, referred to as modified gravity theories, provide an alternative explanation to dark energy for the observed accelerating expansion. As discussed in Chap. 1, several classes of theories exist which generally can be divided into theories which introduce a new scalar degree of freedom to Einstein's equations, e.g. scalar tensor or $f(R)$ theories, and those which modify gravity as a result of the changing dimensionality of space, e.g. braneworld gravity.

In many modified gravity models, the time variation of fundamental constants, such as Newton's gravitational constant, G_N, are naturally present. For example, following Dirac's proposal of the possible cosmological variation of constants to explain large number coincidences in the Universe, many theorists developed self consistent scalar-tensor theories, where the space-time variation of a scalar field can couple to gravity producing a time varying $\tilde{G}$. These 'extended quintessence' models are viable alternatives to Einstein's theory of gravity and give rise to a cosmic expansion that accelerates at late epochs, as required.

Scalar-tensor theories, originally proposed by Jordan Jordan (1949) and Brans and Dicke (1961), are the most widely studied class of modified gravity theories and feature massless scalar fields that couple to the tensor field in Einstein's gravity. These theories are a viable alternative to Einstein's theory of general relativity and have a distinctive feature of a spacetime varying gravitational 'constant'.

Calculations with a mesh to allow spatial variations of the scalar field have shown that, in practice, a broad range of extended quintessence models can be effectively described as a theory which features a time varying Newton's constant (Pettorino and Baccigalupi 2008; Li et al. 2010). The modified gravity model that we discuss in this chapter, which involves a time varying gravitational constant, can be considered as a simple parametrisation of a self consistent modified gravity model. For simplicity, it is common to consider a simple class of models where in Eq. 1.20 in Chap. 1, $F(\varphi, R) = F(\varphi)R/2$. It is the $F(\varphi)$ term which has the effect of introducing a spacetime dependent gravitational constant.

The value of Newton's constant as measured in Cavendish like experiments on terrestrial scales, $\lesssim 1\,\text{m}$, is assumed to be the same on all scales. Although the gravitational constant is the least accurately measured of all the fundamental constants, there are experiments which test G_N on different spatial scales and aim to tightly constrain its variation. For example, solar system scale constraints are obtained in weak field experimental tests, using laser ranging techniques, which measure the distance between the earth and the moon (Williams et al. 1996). If Newton's constant varies over cosmological time scales then the main sequence time of stars in globular clusters will be modified. For example, an increase in G_N shortens the life span of these stars, causing them to burn faster (Teller 1948). degl'Innocenti et al. (1996) constrained the time variation of G_N to be $-35 \times 10^{-12}\mathrm{year}^{-1} \lesssim \dot{G}/G \gtrsim 7 \times 10^{-12}\mathrm{year}^{-1}$, by assuming that the age of globular clusters was between 8–20 Gyear. Another important constraint on the time variation of G_N comes from observing the masses of neutron stars formed at different redshifts. In the late stages of stellar evolution the Fermi pressure of the gas balanced by the strength of gravity determine the Chandrasekhar mass, $\propto G_N^{3/2}$. Assuming that the mean neutron star mass is equal to the Chandrasekhar mass, observations of neutron star binaries can limit the allowed variation of G_N (Thorsett 1996).

If G_N changes during the radiation dominated phase of the Universe's history this will alter the expansion rate during the synthesis of light nuclei in the early Universe, at the epoch of Big Bang nucleosynthesis (BBN), causing freeze out, when nuclear reactions end, to occur at a different time. Taking an explicit form for the evolution of G_N, derived from a scalar tensor theory, it is possible to constrain its variation using observed primordial ^{4}He abundances (Umezu et al. 2005; Clifton et al. 2005). A time-varying G_N will also modify the temperature overdensities measured in the CMB. For example, the CMB peaks shift to larger (smaller) scales with increasing (decreasing) G_N. Fitting to CMB measurements results in a limit on the variation of G of $\dot{G}/G = (-9.6 \sim +8.1) \times 10^{-12}\mathrm{year}^{-1}$, consistent with constraints from BBN and neutron star masses (Chan and Chu 2007).

In extended quintessence cosmologies the background expansion of the universe, which is described by the Friedmann equation, is given by

$$H^2 = \frac{8\pi G_N}{3F(\varphi)}\left(\rho_{\mathrm{fluid}} + \frac{1}{2}\dot{\varphi}^2 + V(\varphi) - 3H\dot{F}\right),\tag{5.3}$$

where the $3H\dot{F}$ term can be omitted as it is negligible (Pettorino and Baccigalupi 2008). The $8\pi G_N/F$ term in this equation modifies the gravitational interaction from that in general relativity and can be parametrized as a spacetime varying gravitational constant.

Pettorino and Baccigalupi (2008) derived the linear perturbations in an extended quintessence cosmology in the Newtonian limit and found that the Poisson equation could be written in the usual way as

$$2k^2\Phi_E = \frac{8\pi G_N}{F}\rho_{\mathrm{m}}\delta_{\mathrm{m}},\tag{5.4}$$

where ρ_{m} and $\delta_{\mathrm{m}} = (\rho_{\mathrm{m}} - \bar{\rho}_{\mathrm{m}})/\bar{\rho}_{\mathrm{m}}$ are the matter density and perturbation respectively and the gravitational potential is re-defined as

$$\Phi_E = \left(1 + \frac{1}{2}\frac{F_{,\varphi}^2}{F + F_{,\varphi}^2}\right)\Phi,\tag{5.5}$$

where $F_{,\varphi}^2$ denotes the derivative of F with respect to φ. This in turn modifies the Euler equation (see Pettorino and Baccigalupi 2008, for details). The resulting

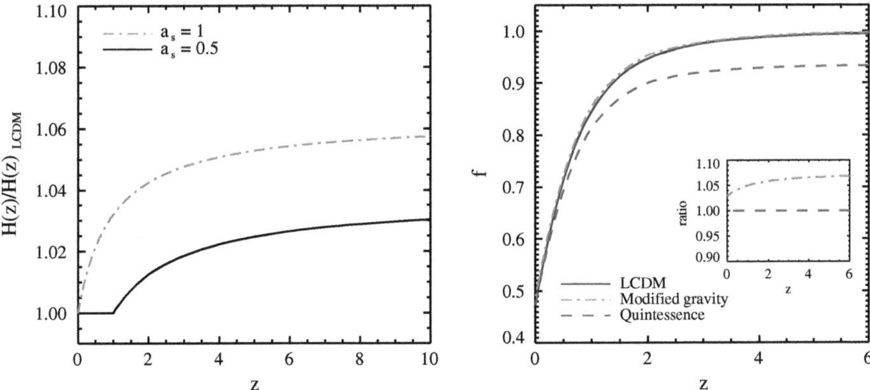

Fig. 5.1 *Left panel* The two different lines show the ratio of the expansion rate, $H(z)$, for two modified gravity models to the expansion rate in ΛCDM. In the *left panel* we plot the expansion history for a modified gravity model with parameters $a_s = 1$ ($a_s = 0.5$), divided by $H(z)$ for ΛCDM, as a (*dot dashed (solid) line*). *Right panel* The linear growth rate, f, as a function of redshift for ΛCDM, a modified gravity cosmology and a quintessence model. In the *right panel*, the linear theory growth rate is plotted as a function of time for a modified gravity model with $a_s = 1$ and $\mu_0^2 = 1.13$ in Eq. 5.8 (*dot dashed line*). The growth rate for a quintessence model, which has the same expansion history as the modified gravity model, is plotted as a *dashed line*. The growth rate for ΛCDM is shown as a *solid line*. The inset panel shows the ratio of f for the modified gravity model to the quintessence model as a function of redshift (*dot dashed line*)

modifications can be expressed in terms of a gravitational constant which is now varying in time as

$$\tilde{G} = \frac{2(F + 2F_{,\varphi}^2)}{(2F + 3F_{,\varphi}^2)} \frac{G_N}{F} \,. \tag{5.6}$$

Cosmological N-body simulations of extended quintessence cosmologies need to account for both the gravitational correction due to a varying G in the Poisson equation and a modified expansion history given in Eq. 5.3. In this chapter we consider a simple model for a time varying Newton's constant (Zahn and Zaldarriaga 2003; Umezu et al. 2005; Chan and Chu 2007),

$$\tilde{G} = \mu^2 G_N \,, \tag{5.7}$$

where

$$\mu^2 = \begin{cases} \mu_0^2 & \text{if } a < a_* \\ 1 - \frac{a_s - a}{a_s - a_*}(1 - \mu_0^2) & \text{if } a_* \leq a \leq a_s \\ 1 & \text{if } a > a_s \,. \end{cases} \tag{5.8}$$

This parametrization describes a smoothly varying $\tilde{G}$ which converges slowly to its present value, G_N and is a more physical model than parametrizations based on step functions which have been considered previously in the literature (Cui et al. 2010). The scale factor, a_*, is taken as the time of photon decoupling and the parameters μ_0 and a_s quantify the deviation of $\tilde{G}$ from the present laboratory measured value, G_N, and the scale factor at which $\tilde{G}$ and G_N are equal, respectively.

We shall assume that this parametrization for $\tilde{G}$ describes a simple extended quintessence model where the coupling $F = 1/\mu^2$ and the background evolution is given by

$$H^2 = H_0^2 \frac{\tilde{G}}{G_N} \left(\frac{\Omega_{\rm m}}{a^2} + \Omega_{\rm DE} e^{3 \int_a^1 {\rm dln}a'[1+w(a')]} \right). \tag{5.9}$$

Note that here we assume an equation of state $w = -1$ in the modified gravity model to match ΛCDM. Here $\Omega_{\rm DE}$ is the ratio of the dark energy density to the critical density today. We assume the Poisson equation is given by Eq. 5.4 with $F = 1/\mu^2$. In the left panel in Fig. 5.1, we plot the ratio of the Hubble rate for two different cosmological models with varying G, to the Hubble rate for a ΛCDM cosmology as a function of redshift. The dot dashed line corresponds to $\tilde{G}$ with $\mu_0^2 = 1.13$ and $a_s = 1$ in Eq. 5.8, while the black solid line uses the parameters $\mu_0^2 = 1.075$ and $a_s = 0.5$. If we require $\tilde{G}$ to converge to G_N at higher redshifts, $z > 0$, then the permitted variation of $\tilde{G}$ from G_N decreases. The maximum deviation of $\tilde{G}$ from G_N which is compatible with CMB measurements occurs for a stabilization redshift corresponding to an expansion factor of $a_s = 1$. We use the parametrization in Eqs. 5.7 and 5.8 for a varying G model with parameters $\mu_0^2 = 1.13$ and $a_s = 1$. We can then construct a quintessence model which has the same expansion history as the modified gravity model. We fit the parameters w_0 and w_a (Linder 2003) in

$$H^2(a) = H_0^2 \left(\frac{\Omega_{\rm m}}{a^3} + \Omega_{\rm DE} e^{-3w_a(1-a)} a^{-3(1+w_0+w_a)} \right) \tag{5.10}$$

to the expansion history for the varying G model using MPFIT (Markwardt 2009). Using tabulated values for w_0 and w_a in the redshift range $z \in [0, 200]$ we were able to reproduce the expansion history of the varying G model to better than 0.25 %. This quintessence model is consistent with current constraints on dynamical dark energy which feature a time varying equation of state (Komatsu et al. 2009).

5.3 Results

5.3.1 Simulation Details

The linear theory power spectrum used to generate the initial conditions was obtained using CAMB (Lewis and Bridle 2002) as in the previous simulations in this thesis. Following previous authors (Laszlo and Bean 2008; Bertschinger and Zukin 2008), we assume that the Jeans length is smaller than the scales of interest at our starting redshift, $z = 200$, and that modified gravity has not yet become important. We therefore assume a ΛCDM cosmology and generate the linear theory power spectrum using CAMB. To obtain errors on our measurements we also ran 10 lower resolution simulations with 512^3 particles in a computational box of comoving length $1500\,h^{-1}$Mpc, each with a different realisation of the density field. The full resolution run has 1024^3 particles in a simulation box of $1500\,h^{-1}$Mpc on a side. The power spectrum was computed by assigning the particles to a mesh using the cloud in cell (CIC) assignment scheme and performing a fast Fourier transform of the density field as carried out in previous chapters. We use a common expression for the fractional error in the power spectrum (Feldman et al. 1994)

$$\frac{\sigma}{P} = \sqrt{\frac{2}{n_{\mathrm{modes}}}} \left(1 + \frac{1}{\bar{n}P}\right), \tag{5.11}$$

where P is the measured power spectrum, $\bar{n}$ is the Poisson shot noise of the simulation and the number of Fourier modes is $n_{\mathrm{modes}} = Vk^2\delta k/(2\pi^2)$, where V is the survey volume. For the initial conditions the linear growth rate for each model and ΛCDM was obtained by solving Eq. 5.2 numerically and is plotted in Fig. 5.1 as a function of redshift. For all the models, we used the following cosmological parameters: $\Omega_{\mathrm{m}} = 0.26, \Omega_{\mathrm{DE}} = 0.74, \Omega_{\mathrm{b}} = 0.044, h_0 = 0.715$ and a spectral index of $n_{\mathrm{S}} = 0.96$ (Sánchez et al. 2009).

In order to test the code used for the modified gravity and quintessence simulations we check that the linear growth of matter in the simulations agrees with the linear theory predictions. In linear theory the power spectrum at redshift z is a scaled version of the power spectrum at an earlier redshift, $\tilde{z}$, according to Eq. 5.2. In the upper and lower panels in Fig. 5.2, we plot the power spectra measured at $z = 0$ (solid), $z = 1$ (dot dashed) and $z = 2$ (dashed) divided by the power spectrum at redshift 5, scaled to take out the difference between the growth factor at $z = 5$ and the redshift plotted in the panel, for the modified gravity and the quintessence model respectively. Using this early redshift power spectrum output at $z = 5$ in the ratio removes the sample variance on large scales and is justified as the density perturbations are still growing according to linear theory at this time. Both models fit the theoretical predictions for their linear growth to a precision of $<0.05\,\%$ on scales $k < 0.01\,h/$Mpc, showing that our modifications to Gadget-2 are accurate.

5.3.2 Redshift Space Distortions

Here we use large volume N-body simulations to carry out the first direct test of the idea that a dark energy cosmology and a modified gravity model which, by construction, have exactly the same expansion history, can be distinguished by a measurement of the rate at which cosmic structure grows. The modified gravity model has a time-varying gravitational constant, $\tilde{G}(a) = \mu^2(a)G_N$, where μ^2 is a

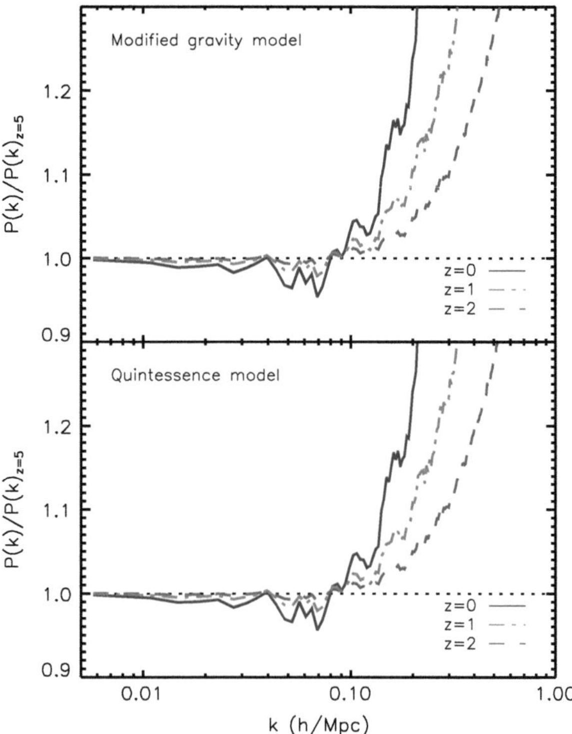

Fig. 5.2 *Top* (*Bottom*) panel the ratio of the modified gravity (quintessence) model power spectrum at three redshifts to the power spectrum at $z = 5$ output from the simulation. The power spectra at each redshift shown have been scaled by the squared ratio of the growth factor at that redshift and the growth factor at $z = 5$ in each cosmology. The ratios at redshift $z = 2$, $z = 1$, and $z = 0$ are shown as *dashed*, *dot dashed* and *solid lines* respectively

linear function of the scale factor a, varying from $\mu^2 = 1.13$ in the early Universe ($a \rightarrow 0$) to $\mu^2(a = 1) = 1$ today, and is consistent with current observational constraints.

The ratio of the quadrupole to monopole moment of the matter power spectrum is plotted in Fig. 5.3 at three output redshifts, $z = 0, 0.5$ and 1. The simulation results show that this ratio has a strong dependence on scale. This can be contrasted with the prediction of linear perturbation theory, (Cole et al. 1994), $P_2^s(k)/P_0^s(k) = (4f/3 + 4f^2/7)/(1 + 2f/3 + f^2/5)$, which is independent of scale (horizontal lines). The quadrupole to monopole ratio increases in amplitude with redshift, due to the associated evolution in the matter density parameter. At $z = 0$ there is a 2.5 % difference between the linear theory growth rates in each model. However, at this level, the measured ratios P_2/P_0 in the two models are indistinguishable on the very largest scales $k < 0.02h/\text{Mpc}$ where our measurements match the linear perturbation theory predictions (green dotted and blue dashed horizontal lines). At $z = 0.5$ and $z = 1$ the linear theory predictions for the growth rates in the two models differ by 4 and 6 % respectively. The error on this ratio measured from the ten lower resolution simulations are shown as a grey shaded region in Fig. 5.3.

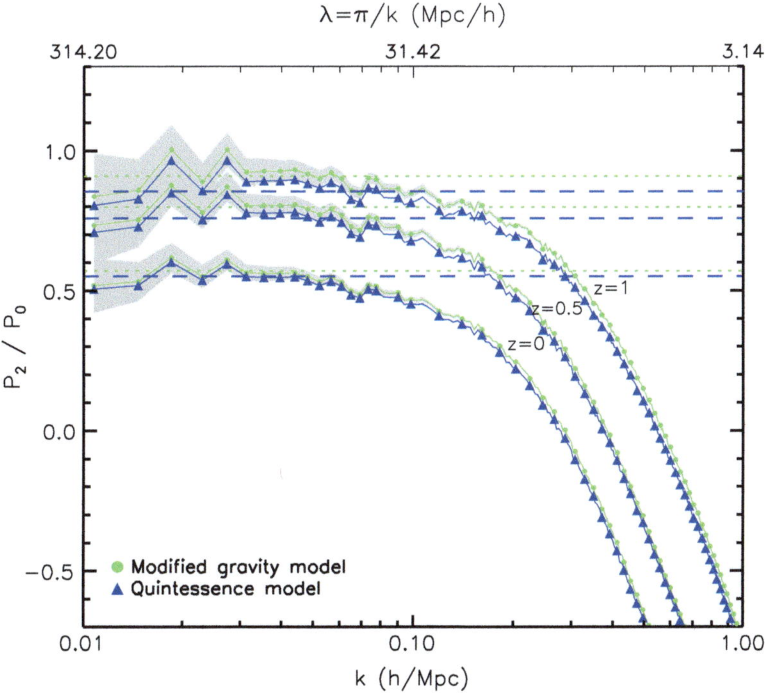

Fig. 5.3 The ratio of the quadrupole to monopole moments of the power spectrum, P_2/P_0 as a function of wavenumber k, where higher values of k correspond to smaller physical scales. The moments are determined in a harmonic analysis of the power spectrum. The points show measurements from the N-body simulations, with *green circles* showing the results from the modified gravity model and *blue triangles* the quintessence model. The shading indicates the error on the ratio, estimated from the scatter over 10 lower resolution simulations. The *horizontal lines* show the predictions of the linear theory model, with the colours having the same meaning as those used for the points. The ratio is shown for three epochs corresponding to redshifts $z = 0$, 0.5 and 1, in order of increasing amplitude. The simulation results show a strong dependence on wavenumber whereas the linear theory predictions are independent of scale

5.3.3 Measuring the Growth Rate

As discussed in Chap. 4, other models have been developed to describe the distortion of the clustering pattern due to peculiar motions, which we now apply to the measurements from the simulations. In addition to the linear theory model described above, we consider two variants. The first is the Gaussian model given in Eq. 4.10 in Chap. 4, in which linear theory is combined with a parametrization for the velocity dispersion on small scales. Here, we refer to this as the "linear theory plus damping" model. The damping introduces a scale dependence into the ratio P_2/P_0. The second model, given in Eq. 4.11 in Chap. 4, takes into account deviations from linear theory, as well as including small scale damping (Scoccimarro 2004; Jennings et al. 2010):

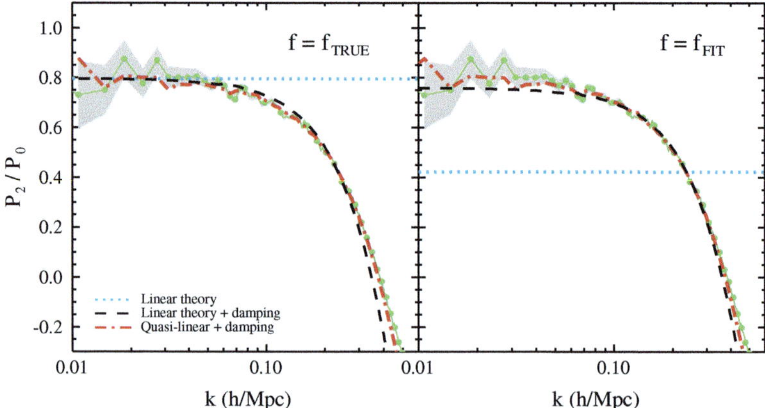

Fig. 5.4 The ratio of the quadrupole to monopole moment of the redshift space power spectrum for the modified gravity cosmology together with three models for P_2^s/P_0^s, using the correct linear growth rate, $f = f_{\mathrm{TRUE}}$ (*left panel*), and the value of f obtained in the χ^2 fit over $0.01 \leq k(h/\mathrm{Mpc})$ ≤ 0.25, $f = f_{\mathrm{FIT}}$ (*right panel*). The P_2^s/P_0^s ratio measured from the high resolution simulation with 1024^3 particles is shown as *green circles*. The error bars (*shaded region*) represent the propagated errors from the ten lower resolution simulations. The quasi-linear plus damping model, the linear theory and the linear theory plus damping model are shown in both panels as *red dot dashed*, *cyan dotted* and *black dashed lines*. In the *left* panel the best fit value for σ_p (σ_v) obtained in the range $0.01 \leq k(h/\mathrm{Mpc}) \leq 0.25$, with fixed f, was used for the linear theory plus damping (quasi-linear plus damping) model

we refer to this as the "quasi-linear theory plus damping" model. We fit these models to the power spectrum measured from the $z = 0.5$ output of our simulations which is one of the target redshifts for the proposed galaxy redshift survey Euclid.

In Fig. 5.4, we plot the measured ratio P_2^s/P_0^s, for the modified gravity cosmology at $z = 0.5$, together with the predictions for this ratio using the quasi-linear plus damping model (red dot dashed line), the linear theory (cyan dotted line) and the linear theory plus damping model (black dashed line). In the left panel, the correct value of f for this cosmology together with the best fit value for σ_p and σ_v in the range $0.01 \leq k(h/\mathrm{Mpc}) \leq 0.25$ was used for the linear theory plus damping and quasi-linear plus damping model respectively. Using another common model for the redshift space power spectrum, the so-called 'dispersion' model (Peacock and Dodds 1994), we found similar values for f to the linear theory plus damping model, when fitting to both P_0^s/P_r and P_2^s/P_0^s. For clarity we have omitted this model from Fig. 5.4. In the right panel, the best fit value for f obtained by fitting over the previous range of scales has been used for all models plotted. It is clear that both the linear theory and the linear theory plus damping models fail to predict the correct value for f, with the best fitting values differing by $\sim 40\%$ and $\sim 6\%$ respectively from the true value. The value of f obtained for the linear model depends on the maximum value of k used in the fit. The linear model plotted in the right panel in Fig. 5.4 uses the value

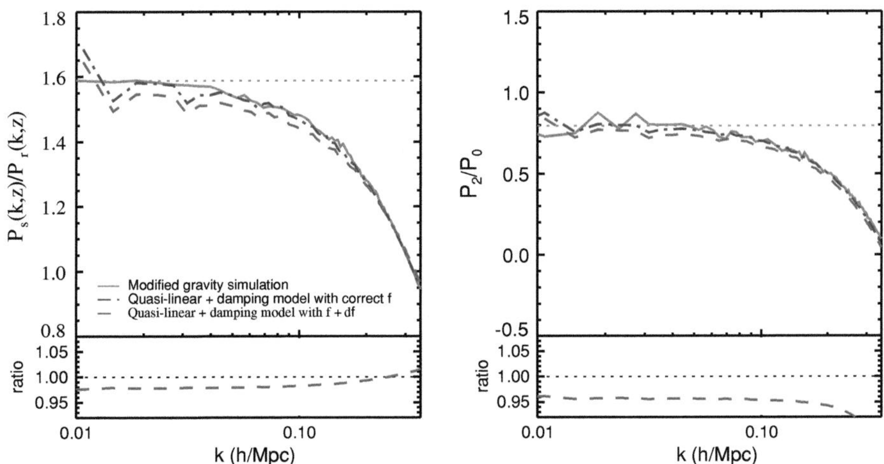

Fig. 5.5 *Left* panel: The ratio of the monopole of the redshift space power spectrum to the real space power spectrum at $z = 0.5$, as a function of wavenumber, k(h/Mpc). *Right* panel: The ratio of the quadrupole to monopole moment of the redshift space $P(k)$ at $z = 0.5$, as a function of wavenumber. Both panels show the quasi-linear plus damping model for the ratios using different values for the linear growth rate, f. In both panels we plot the quasi-linear plus damping model using $f = f_{\text{TRUE}}$ (*dot dashed*) and $f = 1.05 f_{\text{TRUE}}$ (*dashed*), using the best fit value for σ_v in the range $0.01 \leq k$(h/Mpc) ≤ 0.25. The *dashed line* in the bottom left and right panels show the ratio of the quasi-linear plus damping model using $f = 1.05 f_{\text{TRUE}}$ to the same model using $f = f_{\text{TRUE}}$ for P_2^s / P_0^s and P_0^s / P_r respectively

of f recovered when $k_{\text{max}} = 0.25$. The quasi-linear plus damping model recovers the correct value of f in this range to a precision of $\sim 0.64\%$.

In Fig. 5.5, we plot the ratios P_0^s / P_r and P_2^s / P_0^s for the modified gravity model, at $z = 0.5$, in the left and right panels respectively. The quasi-linear plus damping model with the correct value of the linear growth rate, $f = f_{\text{TRUE}}$, is plotted as a dot dashed line. The dashed line shows the quasi-linear plus damping model with a linear growth rate which differs by 5 % from the true value, f_{TRUE}. In the lower left (right) panel we plot the ratio of the quasi-linear plus damping model using $f = 1.05 f_{\text{TRUE}}$ to the same model with $f = f_{\text{TRUE}}$ for the P_0^s / P_r (P_2^s / P_0^s) ratio as a dashed line. Changing f by 5 % produces a $\sim 2\%$ change in the quasi-linear plus damping model for the P_0^s / P_r ratio but a larger, $\sim 4\%$, change in the P_2^s / P_0^s ratio.

To test these models for the redshift space power spectrum further we vary the maximum wavenumber, k_{max}, used in the fit and plot the recovered growth rate as a function of k_{max} in Fig. 5.6. If we had an accurate model of P_2 / P_0, we would recover the correct value for the growth rate f and the answer would be independent of the value of k_{max} adopted, with the only change being the error on the growth rate. Fig. 5.6 shows that the quasi-linear plus damping model comes closest to meeting this ideal. This model breaks down beyond $k_{\text{max}} \sim 0.3\, h\text{Mpc}^{-1}$, which suggests that the modelling of the small scale velocity dispersion needs to be improved. Most

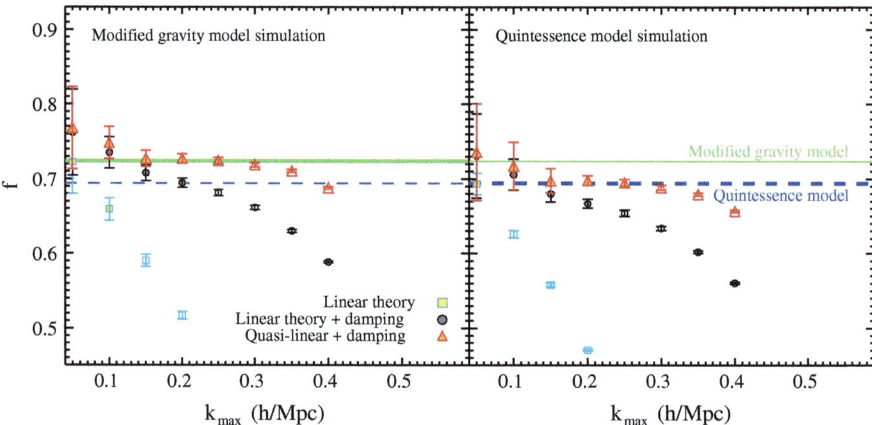

Fig. 5.6 Measurements of the linear growth rate of cosmic structure, f. The results are plotted as a function of the maximum wavenumber, $k_{\max}$ (h/Mpc), used in the fit. The different symbols show the results of fitting to P_2/P_0 at $z = 0.5$ (see Fig. 5.3) using different models: linear theory—*squares*, linear theory plus damping—*circles*, quasi-linear plus damping—*triangles*. The symbols are filled in for scales over which the model is a good description of the measured ratio. The error bars represent the 1σ uncertainty in the fit. In the *left* hand panel we fit to the modified gravity model and aim to recover the true growth factor shown by the thick *green horizontal line*. In the *right* panel we fit to the ratio measured from the quintessence model, in which case the target growth factor is shown by the thick *blue dashed line*. The quasi-linear plus damping model performs best, recovering the correct growth factor in each case over the widest range of wavenumbers. This model is an accurate model of the simulation results up to $k_{\max} = 0.3$h/Mpc. The linear and linear plus damping models are less successful, and only recover the correct answer over a very limited range of wavenumbers. Their application over a wider range of scales would lead to a systematic error in the growth factor similar to or larger than the difference in the growth factors between the two models

importantly, this model recovers the correct value for f and can distinguish between the two cosmologies. The models based on linear theory perform less well. In fact, the answer depends strongly on the maximum wavenumber retained in the fit. In Fig. 5.6 the symbols are filled in for scales over which the model is a good description of the measured ratio. We consider a model as being a good description of the data if $\chi^2/\nu \sim 1$, where ν is the number of degrees of freedom. As shown in Fig. 5.3, the expression for P_2^s/P_0^s in this model is more sensitive to changes in f and as a result the 1 sigma error bars for f in Fig. 5.6 are smaller when fitting to P_2^s/P_0^s compared to P_0^s/P_r.

5.4 Summary

The next generation of galaxy redshift surveys aim to resolve some of the fundamental questions in modern cosmology, such as whether general relativity needs to be modified or if a dark energy component is driving the accelerating expansion.

We have measured the redshift space distortions from two simulations with different cosmologies and demonstrated that a modified gravity model, described by a time varying Newton's constant, and a dark energy model, which have identical expansion histories, have measurably different growth rates. We have tested several models for redshift space distortions of clustering including the commonly used linear theory and linear theory plus Gaussian damping models. We find that these two models fail to recover the correct value for the growth rate. However, a quasi-linear model which includes non-linear velocity divergence terms is far more accurate and would allow us to distinguish between these two competing cosmologies.

Even though the scales we consider are large it is clear that there are important departures from linear theory which can only be modelled accurately using an N-body simulation (Jennings et al. 2010). There is a real chance that without such guidance from a simulation, the application of the linear theory or linear theory plus damping models could lead to systematic errors of the same order as the difference in f between the two competing cosmologies. In this event, these models would give the wrong conclusion about the physics driving the cosmic acceleration. Our results indicate that by using an improved model for the power spectrum in redshift space to constrain the linear growth rate, together with an accurate measurement of the expansion history, we will be able to identify variations in Newton's gravitational constant.

References

Bertschinger E, Zukin P (2008) Phys Rev D 78:024015

Brans C, Dicke RH (1961) Phys Rev 124:925

Chan KC, Chu M (2007) Phys Rev D 75:083521

Clifton T, Barrow JD, Scherrer RJ (2005) Phys Rev D 71:123526

Cole S, Fisher KB, Weinberg DH (1994) Mon Notices Royal Astron Soc 267:785

Cui W, Zhang P, Yang X (2010) Phys Rev D 81:103528

degl'Innocenti S, Fiorentini G, Raffelt GG, Ricci B, Weiss A (1996) A&A, 312:345

Feldman HA, Kaiser N, Peacock JA (1994) Astrophys J 426:23

Jennings E, Baugh CM, Pascoli S (2010) Mon Notices Royal Astron Soc 1572

Jordan P (1949) Nature 164:637

Komatsu E et al (2009) Astrophys J Suppl Ser 180:330

Laszlo I, Bean R (2008) Phys Rev D 77:024048

Lewis A, Bridle S (2002) Phys Rev D 66:103511

Li B, Mota DF, Barrow JD (2010) ArXiv e-prints

Linder EV (2003) Phys Rev Lett 90:091301

Linder EV (2005) Phys Rev D 72:043529

Lue A, Scoccimarro R, Starkman GD (2004) Phys Rev D 69:124015

Markwardt CB (2009) Astron Soc Pac Conf Ser 411:251

Peacock JA, Dodds SJ (1994) Mon Notices Royal Astron Soc 267:1020

Pettorino V, Baccigalupi C (2008) Phys Rev D 77:103003

Sánchez AG, Crocce M, Cabré A, Baugh CM, Gaztañaga E (2009) Mon Notices Royal Astron Soc 400:1643

Scoccimarro R (2004) Phys Rev D 70:083007
Teller E (1948) Phys Rev 73:801
Thorsett SE (1996) Phys Rev Lett 77:1432
Umezu K, Ichiki K, Yahiro M (2005) Phys Rev D 72:044010
Williams JG, Newhall XX, Dickey JO (1996) Phys Rev D 53:6730
Zahn O, Zaldarriaga M (2003) Phys Rev D 67:063002

Chapter 6
Conclusions

The current evidence for the accelerating cosmic expansion is substantial, with a host of different observations suggesting that dark energy makes up $\sim 70\,\%$ of our Universe. The physics driving this acceleration is still unknown and represents the most compelling and challenging question to be answered in our standard cosmological model. The most popular current explanation is the cosmological constant, the vacuum energy of space, which is a negative pressure dark energy component giving rise to a homogeneous expansion of the Universe. Explaining the observed value of the cosmological constant is a serious challenge and will require new physics beyond the standard model of particle physics and cosmology. Several other candidate theories such as dynamical dark energy e.g. quintessence, or modified gravity models exist and can fit the current data as well as, and in some cases better than, the concordance ΛCDM model (Dantas et al. 2010). The exciting prospect is that some of these models may leave detectable signatures on the growth rate and in the distribution of large scale structure in the Universe, allowing us to distinguish them from ΛCDM.

As discussed in Chap. 1, there are several observational probes which can be used to constrain the properties of dark energy and modified gravity. These observations, such as measurements of Type Ia SN light curves and the temperature power spectrum of the CMB, are sensitive to different physical processes at different epochs and provide powerful constraints on cosmology when combined together. Broadly speaking these observations can be divided into those that measure the expansion rate and geometry of the Universe e.g. Type Ia SN, BAO and CMB measurements, the growth of structure e.g. redshift space distortions in the power spectrum, or a combination of the two e.g. weak lensing and cluster mass functions. Measurements of the expansion history can constrain the dark energy equation of state and its variation both today and at high redshift. A definite detection of $w \neq -1$ would rule out ΛCDM but it would leave us with a host of viable dynamical dark energy and modified gravity models. However a dark energy or modified theory with identical expansion histories will have different growth rates for structure in our Universe, allowing us to distinguish these two models and break this degeneracy.

E. Jennings, *Simulations of Dark Energy Cosmologies*, Springer Theses, 97
DOI: 10.1007/978-3-642-29339-9_6, © Springer-Verlag Berlin Heidelberg 2012

Cosmological N-body simulations are the theorist's tool of choice for modelling the final stages of perturbation collapse. To date, the overwhelming majority of simulations have used the concordance ΛCDM cosmology. Here we simulate different dark energy models and study their observational signatures. A small number of papers have used N-body simulations to test scalar field cosmologies by modifying the expansion history alone (Ma et al. 1999; Linder and Jenkins 2003; Klypin et al. 2003; Francis et al. 2008; Grossi and Springel 2009; Alimi et al. 2009. In Chap. 3 we carried out the most realistic simulations of quintessence dark energy to date, using an accurate parametrization for the quintessence models equation of state and a consistent linear theory power spectrum and appropriate cosmological parameters (so that the various models match the CMB, BAO and SN constraints). We found that these models can have a significant impact on the growth of structure when correctly simulated. By measuring the abundance of dark matter halos we provided theoretical predictions for these dark energy models which will distinguish them from the standard cosmological model in future galaxy surveys. For example, a number of optical imaging surveys such as the one proposed with the LSST, plan to study the properties of dark energy through the large scale distribution of matter over a wide redshift interval (LSST Science Collaborations et al. 2009). The LSST will observe $\sim 10^{10}$ galaxies over 20,000 deg^2 and will be able to measure the abundance of clusters and the BAO as a function of redshift. When combined with weak lensing shear-shear correlations these measurements will constrain Ω_{DE} with an error of 0.003 with uncertainties on w_0 and w_a of 0.03 and 0.1 respectively (Albrecht and Bernstein 2007; Fang and Haiman 2007). The LSST will also be able to detect scales well beyond the turnover in the power spectrum which will provide powerful constraints on models with non-negligible amounts of dark energy at high redshifts which alters the shape of the turnover, such as the CNR and the AS model considered in this thesis.

With this level of precision anticipated from future surveys such as the LSST, the work presented here represents a significant step forward in simulating quintessence dark energy. Overall, our analysis shows that the prospects for detecting dynamical dark energy, which features a late time transition, using the halo mass function at $z > 0$ are good, provided a good proxy can be found for mass. Parameter degeneracies allow some quintessence models to have identical BAO peak positions to ΛCDM and so these measurements alone will not be able to rule out some quintessence models. Although including the dark energy perturbations has been found to increase these degeneracies (Weller and Lewis 2003), incorporating them into the N-body code would clearly be the next step towards simulating quintessential dark matter with a full physical model. Although in many quintessence models the dark energy clusters on very large scales today ($k < 0.02\,\mathrm{hMpc}^{-1}$) (Weller and Lewis 2003) and the perturbations are generally small ($\delta_{\mathrm{DE}} \sim 10^{-1}$), these perturbations may nevertheless have some impact on the dark matter structure in a full N-body simulation of the nonlinear growth (Li et al. 2010).

In Chap. 4, we measured the power spectrum in redshift space from several quintessence dark energy simulations, which is anisotropic due to peculiar velocities which distort the clustering signal on all scales. Modelling the redshift space distortions in either the matter power spectrum or the correlation function allows us to measure the growth rate of structure which is a crucial test of general relativity and the physics driving the accelerating expansion. We demonstrate that the linear theory prediction for the power spectrum in redshift space is a poor fit to the measured distortions, even on surprisingly large scales $k \sim 0.05\,\mathrm{h/Mpc}$. We also consider an improved model for the redshift space distortions which accounts for velocity divergence non-linearities. From our results, it is clear that including the non-linear velocity divergence terms results in an improved model on scales $k < 0.2\,\mathrm{h/Mpc}$ for different cosmological models. Using a density-velocity relation, we provide a cosmology independent formula for generating the velocity power spectrum from the non-linear matter power spectrum. These results are timely and will be relevant for future galaxy redshift surveys such as Euclid and BigBOSS (Cimatti et al. 2009; Schlegel et al. 2007). Current galaxy redshift surveys can provide only very weak constraints on $P_{\delta\theta}$ and $P_{\theta\theta}$ (Tegmark et al. 2002) but both BigBOSS and Euclid plan to map the galaxy distribution at higher redshifts and to a greater precision than previously possible. The relation given in this thesis between the non-linear velocity divergence and matter power spectra will be useful for analysing redshift space distortions in future galaxy surveys as it removes the need to use noisier and sparser velocity data.

In addition to the many dark energy models considered which can explain the accelerating expansion, it may be that an even more radical solution is needed such as modifying general relativity itself. At present there are two key probes, gravitational lensing and the growth rate of structure, which will allow us to test general relativity. Modifications to general relativity can be parametrized using two variables which can vary in space and time-one is an effective gravitational constant which describes any deviations from Newton's constant and the other is the so called slip parameter, $\zeta = \Psi/\Phi$, which describes the difference between the gravitational potentials, Ψ and Φ (see e.g. Bertschinger and Zukin 2008). In Chap. 5 we measure the redshift space distortions from two simulations with different cosmologies and demonstrate that a modified gravity model, described by a time varying Newton's constant, and a dark energy model, which have identical expansion histories, have measurably different growth rates. We test several models for the redshift space distortions including the commonly used linear theory and Gaussian models. We find that these two models fail to recover the correct value for the growth rate, while a quasi-linear model which includes non-linear velocity divergence terms, discussed in Chap. 4, is far more accurate and would allow us to distinguish these two competing cosmologies.

The next generation of galaxy redshift surveys aim to resolve some of the fundamental questions in modern cosmology and reveal if general relativity is the correct description of gravity or if a dark energy component is driving the accelerating expansion. For example, ESA's Euclid mission (Cimatti et al. 2009) which plans to survey $20,000\,\mathrm{deg}^2$ with $\Delta z = 0.1$, corresponding to a volume of $\sim 1\,\mathrm{Gpc}^3$ at $z = 0.1$, will be able to constrain the linear growth rate to $<2\,\%$ and the dark energy equation of state parameters w_0 and w_a to 2 and $10\,\%$ respectively. Our results indicate that using a correct model for the power spectrum in redshift space, a constraint on the linear growth rate of better than 2% for several redshift bins from $z = 0$ to $z = 2$, together with an accurate measurement of the expansion history to $<4\%$ would identify variations in Newton's gravitational constant, providing a strong signal that modified gravity describes our Universe.

There are many ways in which future research on dark energy and modified gravity cosmologies can benefit from the use of N-body simulations. As a first example, at present only one consistent modified gravity model has ever been tested using a simulation which lacked spatial resolution (Oyaizu et al. 2008). Clearly more work is needed to add modified gravity models into simulations and to test the impact these models have on the measured redshift space distortions and the determination of the growth rate. These modifications could be accounted for using parametrizations for a variable gravitational constant, $G(t, \vec{x})$ and varying gravitational potentials, $\zeta(t, \vec{x}) = \Psi/\Phi$, which can depend on space and time. Secondly, another issue at present in determining the growth rate is whether or not to use the two point correlation function instead of its Fourier transform, the power spectrum. Previous measurements of the correlation function in redshift space (Guzzo et al. 2008) have found agreement with linear theory predictions on very small scales, $r \sim 20\,h^{-1}\mathrm{Mpc}$. These clustering measurements in real space are obtained by integrating over all the Fourier modes and it is not clear what impact this has on the resulting errors. The correspondence between the correlation function and the power spectrum errors can be investigated with accurate simulations of different cosmologies and would significantly improve our current models of redshift space distortions and future constraints on the growth rate.

References

Albrecht A, Bernstein G (2007) Phys. Rev. D 75:103003
Alimi J, Fuzfa A, Boucher V, Rasera Y, Courtin J, Corasaniti P (2009) arXiv:0903.5490.
Bertschinger E, Zukin P (2008) Phys. Rev. D 78:024015
Cimatti A et al (2009) Exp Astron 23:39
Dantas MA, Alcaniz JS, Mania D, Ratra B (2010) arXiv e-prints.
Fang W, Haiman Z (2007) Phys. Rev. D 75:043010
Francis MJ, Lewis GF, Linder EV (2008) MNRAS 394:605
Grossi M, Springel V (2009) MNRAS 394:1559
Guzzo L et al (2008) Nature 451:541
Klypin A, Maccio AV, Mainini R, Bonometto SA (2003) ApJ 599:31
Li B, Mota DF, Barrow JD (2010) arXiv e-prints.
Linder EV, Jenkins A (2003) MNRAS 346:573
LSST Science Collaborations Abell PA, Allison J, Anderson SF, Andrew JR, Angel JRP, Armus L, Arnett D, Asztalos SJ, Axelrod TS et al (2009) arXiv e-prints.
Ma C-P, Caldwell RR, Bode P, Wang L-M (1999) ApJ 521:L1
Oyaizu H, Lima M, Hu W (2008) Phys. Rev. D 78:123524
Schlegel DJ et al (2007) in Bulletin of the American Astronomical Society vol 38, SDSS-III: The Baryon Oscillation Spectroscopic Survey (BOSS). p 966
Tegmark M, Hamilton AJS, Xu Y (2002) MNRAS 335:887
Weller J, Lewis AM (2003) MNRAS 346:987

Appendix A
WMAP Distance Priors

The method suggested in Komatsu et al. (2009) employs three distance priors from measurements of the CMB together with the 'UNION' supernova samples (Kowalski et al. 2008) and the baryon acoustic oscillations (BAO) in the distribution of galaxies (Percival et al. 2007) to explore the best fit parameters for the dynamical dark energy models. In Sects. 3.3.1 and 3.3.2, all of the quintessence simulations were run using the best fit cosmological parameters assuming a ΛCDM model. While this is useful for isolating the effect of the different expansion histories on the growth of structure, this does not yield quintessence models which would automatically satisfy the constraints on distance measurements. Using CMB, supernovae and BAO data in this way is very useful for testing and perhaps even ruling out some of the dark energy quintessence models. In Sect. 3.3.3 we consider the impact of using these new cosmological parameters on the non-linear growth of structure.

These distance priors are derived parameters which depend on the assumed cosmological model and yield constraints on dark energy parameters which are slightly weaker than a full Markov Chain Monte Carlo (MCMC) calculation, as only part of the full WMAP data is used i.e. the C_l spectrum is condensed into 2 or 3 numbers describing peak position and ratios and the polarisation data are ignored. The assumed model is a standard FLRW universe with an effective number of neutrinos equal to 3.04 and a nearly power law primordial power spectrum with negligible primordial gravity waves and entropy fluctuations. These WMAP distance priors are extremely useful for providing cosmological parameter constraints at a reduced computational cost compared to a full MCMC calculation. We shall briefly review the distance scales used in this thesis and the method for finding the best fit parameters for the dark energy models. From measurements of the peaks and troughs of the acoustic oscillations in the photon-baryon plasma in the CMB it is possible to measure two distance ratios (Komatsu et al. 2009). The first ratio is quantified by the 'acoustic scale', l_A, which is defined in terms of the sound horizon at decoupling, $r_s(z_*)$ and the angular diameter distance to the last scattering surface, $D_A(z_*)$, as

E. Jennings, *Simulations of Dark Energy Cosmologies*, Springer Theses,
DOI: 10.1007/978-3-642-29339-9, © Springer-Verlag Berlin Heidelberg 2012

$$l_A = (1 + z_*) \frac{\pi D_A(z_*)}{r_s(z_*)}. \tag{A.1.1}$$

Assuming a flat universe, the proper angular diameter distance is defined as

$$D_A(z) = \frac{c}{(1 + z)} \int_0^z \frac{dz'}{H(z')}, \tag{A.1.2}$$

and the comoving sound horizon is given by

$$r_s(z) = \frac{c}{\sqrt{3}} \int_0^{1/(1+z)} \frac{da}{a^2 H(a) \sqrt{1 + (3\Omega_b/4\Omega_\gamma)a}} \tag{A.1.3}$$

where $\Omega_\gamma = 2.469 \times 10^{-5} h^{-2}$ for $T_{CMB} = 2.725$K (Komatsu et al. 2009) and Ω_b is the ratio of the baryon energy density to the critical density. We shall use the fitting formula proposed by Hu and Sugiyama (1996) for the decoupling epoch z_* which is a function of $\Omega_b h^2$ and $\Omega_m h^2$ only. The second distance ratio measured by the CMB is called the 'shift parameter' (Bond et al. 1997). This is the ratio of the angular diameter distance and the Hubble horizon size at the decoupling epoch which is written as

$$R(z_*) = \frac{\sqrt{\Omega_m H_0^2}}{c} (1 + z_*) D_A(z_*). \tag{A.1.4}$$

Equation (A.1.4) assumes a standard radiation and matter dominated epoch when calculating the sound horizon. The expression for the shift parameter will be modified for quintessence models of dark energy. The proper expression for the shift parameter is given by (Kowalski et al. 2008)

$$R(z_*) = R_{std}(z_*) \left(\int_{z_*}^\infty \frac{dz/\sqrt{\Omega_m(1 + z)^3}}{\int_{z_*}^\infty dz H_0/H(z)} \right), \tag{A.1.5}$$

where R_{std} is the standard shift parameter given in Eq. (A.1.4). This correction to the shift parameter can be substantial for quintessence models with non-negligible amounts of dark energy at early times and so we include this correction for all of the scalar field models in this thesis. The 5-year WMAP constraints on l_A, R and the redshift at decoupling z_* are the WMAP distance priors used to test models of dark energy (Komatsu et al. 2009).

The angular diameter distance at the decoupling epoch can be determined from measurements of the acoustic oscillations in the CMB. These baryon acoustic oscillations are also imprinted on the distribution of matter. Using galaxies as tracers for the underlying matter distribution the clustering perpendicular to the line of sight gives a measurement of the angular diameter distance, $D_A(z)$. BAO data also allow us to measure the expansion rate of the universe, $H(z)$, from observations of clustering along the line of sight. However, current data do not provide us with a measure of $D_A(z)$ and $H(z)$ individually (Gaztanaga et al. 2008).

Using a spherically averaged correlation function to reveal the BAO signal results in an effective distance measure given by (Eisenstein et al. 2005)

$$D_V(z) = \left((1+z)^2 D_A^2(z) \frac{cz}{H(z)} \right)^{1/3}.$$

(A.1.6)

It is the ratio of $D_V(z)$ to the sound horizon, r_s, at the drag epoch, z_{drag}, which determines the peak positions of the BAO signal. The drag epoch is the redshift at which baryons are separated from photons and is slightly later than the decoupling epoch, z_*. Percival et al. (2007) provide $r_s(z_d)/D_V(z)$ at two redshifts, $z = 0.2$ and $z = 0.35$, taken from the Sloan Digital Sky Survey (SDSS) and Two Degree Field Galaxy Redshift Survey (2dFGRS). The two values are $r_s(z_d)/D_V(0.2) = 0.198 \pm 0.0058$ and $r_s(z_d)/D_V(0.35) = 0.1094 \pm 0.0033$.

The UNION supernovae compilation (Kowalski et al. 2008) consists of 307 low redshift SN all processed using the SALT light curve fitter (Guy et al. 2005). This compilation includes older data sets from the Supernova Legacy Survey and ESSENCE Survey as well as a recent dataset observed with HST. Type Ia SN data is extremely useful in breaking parameter degeneracies such as the w, Ω_{DE} degeneracy in the CMB data. A wide range of these two parameters can produce similar angular diameter distances at the redshift of decoupling and so SN constraints, which are almost orthogonal to CMB constraints, help to reduce this parameter space. The current SN data cover a wide range of redshift, $0.02 \leq z \leq 1.7$, but is only able to weakly constrain a dynamical dark energy equation of state, w, at $z \geq 1$. Also, due to a degeneracy with Ω_m, the current SN data by themselves are not able to tightly constrain the present value of w and including measurements involving Ω_m such as CMB or BAO observations break this degeneracy.

Following the prescription of Komatsu et al. (2009) for using the WMAP distance priors it is necessary to find the vector $\vec{x} = (l_A, R, z_*)$ for each quintessence model in order to compute the likelihood, $\mathscr{L}$, as $\chi^2 = -2\ln\mathscr{L} = (x_i - d_i)C_{ij}^{-1}(x_j - d_j)$, where $\vec{d} = (l_A^{WMAP}, R^{WMAP}, z_*^{WMAP})$ and C_{ij}^{-1} is the inverse covariance matrix for the WMAP distance priors. In order to find the best fit cosmological parameters for each quintessence model we minimise the function $\chi^2_{total} = \chi^2_{WMAP} + \chi^2_{BAO} + \chi^2_{SN}$ with respect to $\Omega_m h^2$, $\Omega_b h^2$ and H_0. In appendix D of Komatsu et al. (2009) it can be seen that including the systematic errors has a very small effect on the ΛCDM parameters but can have a significant effect on dark energy parameters. Using a two parameter equation of state for the dark energy Komatsu et al. (2009) found that the parameter constraints weakened considerably after including systematic errors. In calculating χ^2_{SN} in this thesis we have used the covariance matrix for the errors on the SN distance moduli without systematics.

Table A.1 shows the WMAP distance priors computed for each dark energy model using the cosmological parameters from Sánchez et al. (2009). The BAO scale and drag redshift, z_d, are given in Table A.2 using the same parameters. From these tables it is clear that some quintessence models with ΛCDM

Table A.1 WMAP distance priors (Komatsu et al. 2009) for each quintessence model using $\Omega_m h^2$, $\Omega_b h^2$ and H_0 parameters from Sánchez et al. (2009). These parameters were derived assuming a ΛCDM cosmology. $l_A(z_*)$ is the acoustic scale at the epoch of decoupling, z_* and $R(z_*)$ is the shift parameter. $\chi^2_{total} = \chi^2_{WMAP+SN+BAO}$ and ν is the number of degrees of freedom

	$z*$	$l_A(z*)$	$R(z*)$	χ^2_{total}/ν
WMAP 5-yr ML	1090.51 ±0.95	302.10 ± 0.86	1.710 ± 0.019	0
LCDM	1090.65	303.73	1.73	1.09
INV1	–	261.05	1.49	15.34
INV2	–	294.34	1.67	1.81
SUGRA	–	284.03	1.62	3.88
2EXP	–	303.85	1.74	1.09
AS	–	289.69	1.74	2.04
CNR	–	306.71	1.79	1.37

Table A.2 BAO distance measurements (Percival et al. 2007) for each quintessence model using $\Omega_m h^2$, $\Omega_b h^2$ and H_0 parameters from Sánchez et al. (2009). These parameters were derived assuming a ΛCDM cosmology. A fitting formula proposed by Eisenstein and Hu (1998) was used for the drag redshift z_{drag}

	z_{drag}	$r_s(z_{drag})$	$r_s(z_{drag})/D_v(z=0.2)$	$r_s(z_{drag})/D_v(z=0.2)$
WMAP 5-yr	1020.5 ± 1.6	153.3 ± 2.0 Mpc	-	-
Percival et al. (2007)	–	154.758	0.198 ± 0.0058	0.1094 ± 0.0033
LCDM	1020.505	152.68	0.193	0.116
INV1	–	152.534	0.208	0.130
INV2	–	152.682	0.198	0.121
SUGRA	–	152.466	0.198	0.121
2EXP	–	152.003	0.192	0.115
AS	–	143.874	0.183	0.111
CNR	–	150.738	0.191	0.114

cosmological parameters fail to agree with the distance measurements within the current constraints.

With the assumption that $\Omega_m h^2$, $\Omega_b h^2$ and H_0 are tightly constrained by WMAP, BAO and SN data, and as a result their posterior distribution is close to a normal distribution, minimising $\chi^2_{total} = \chi^2_{WMAP} + \chi^2_{BAO} + \chi^2_{SN}$ with respect to these three parameters will be the same as marginalising the posterior distribution. We have fixed the dark energy equation of state parameters for each quintessence model and the 68.3 % confidence intervals for each parameter from minimising χ^2_{total} are shown in Table A.3. The final column in this table is χ^2/ν where ν is the number of degrees of freedom. From Table A.3 it is clear that the INV1 model is unable to fit the data and has a poor $\chi^2/\nu = 2.27$ statistic. Most of the quintessence models favour a lower $\Omega_m h^2$ compared to ΛCDM in order to fit the distance data. As can be seen from Table A.3 the confidence intervals on the three fitted parameters $\Omega_m h^2$, $\Omega_b h^2$ and H_0 are quite large. Once the best fit parameters from

Table A.3 Best fit values for $\Omega_m h^2$, $\Omega_b h^2$ and H_0 with 68.3% confidence intervals from minimising $\chi^2_{total} = \chi^2_{WMAP+SN+BAO}$ for each quintessence model. wCDM WMAP 5-year are the parameter constraints assuming a dynamical dark energy model (Komatsu et al. 2009)

	$10^2\Omega_b h^2$	H_0 (km/s/Mpc)	$\Omega_m h^2$	χ^2_{total}/ν
LCDM WMAP 5-yr Mean	2.267 $^{+0.058}_{-0.059}$	70.5 ± 1.3	0.1358 $^{+0.0037}_{-0.0036}$	
wCDM WMAP 5-yr Mean	2.27 ± 0.06	69.7 ± 1.4	0.1351 ± 0.0051	
Sánchez et al. (2009)	2.267 $^{+0.049}_{-0.05}$	71.5 ± 1.1	0.13343 ± 0.0026	1.09
INV1	3.78 ± 0.145	63.13 ± 0.54	0.1152 ± 0.0103	2.27
INV2	2.35 ± 0.094	68.21 ± 0.70	0.124 ± 0.0065	1.07
SUGRA	2.68 ± 0.105	67.625 ± 0.71	0.1112 ± 0.0075	1.25
2EXP	2.22 ± 0.115	70.01 ± 0.8	0.1386 ± 0.00315	1.05
AS	2.12 ± 0.121	70.42 ± 0.98	0.086 ± 0.0121	1.07
CNR	2.09 ± 0.185	70.05 ± 1.25	0.140 ± 0.0133	1.12

Table A.4 WMAP distance priors (Komatsu et al. 2009) for each quintessence model using the best fit parameters $\Omega_m h^2$, $\Omega_b h^2$ and H_0 given in Table 3.3

	$z*$	$l_A(z*)$	$R(z*)$
LCDM WMAP 5-yr ML	1090.51 ± 0.95	302.10 ± 0.86	1.710 ± 0.019
Sanchez et al. 2009	1090.12 ± 0.93	301.58 ± 0.67	1.701 ± 0.018
INV1	1076.178	292.544	1.519
INV2	1088.716	301.693	1.676
SUGRA	1083.96	298.512	1.596
2EXP	1091.75	302.916	1.749
AS	1087.98	300.237	1.684
CNR	1093.97	303.515	1.809

Table A.5 BAO distance measurements (Percival et al. 2007) for each quintessence model using the best fit parameters $\Omega_m h^2$, $\Omega_b h^2$ and H_0 given in Table 3.3

	z_{drag}	$r_s(z_{drag})$	$r_s/D_V(z=0.2)$	$r_s/D_V(z=0.35)$
WMAP 5-yr	1020.5 ± 1.6	153.3 ± 2.0 Mpc	-	-
Percival et al. (2007)	−	154.758	0.198 ± 0.0058	0.1094 ± 0.0033
INV1	1045.140	146.259	0.1765	0.1103
INV2	1021.192	154.946	0.1921	0.1167
SUGRA	1026.379	155.803	0.1908	0.1161
2EXP	1019.995	150.983	0.1879	0.1123
AS	1010.479	157.745	0.1947	0.1161
CNR	1017.073	150.597	0.1876	0.1128

Table A.3 are used, all of the quintessence models apart from INV1 which we rule out, produce a better fit to the data, as seen in Tables A.5 and A.6, for the WMAP distance priors and the BAO distance measures respectively. As we noted earlier the WMAP distance priors do not contain all of the WMAP power spectrum data and only use the information from the oscillations present at small angular scale (high multipole moments). Neglecting the Integrated-Sachs-Wolfe (ISW) effect at large angular scales (small multipole moments) as well as polarisation data lead to weaker constraints on cosmological parameters in these dark energy models. We have not considered how these distance priors would change with the inclusion of dark energy perturbations (Li et al. 2008). These results are in agreement with previous work fitting cosmological parameters of quintessence models using WMAP first year CMB data and SN data (Corasaniti et al. 2004).

Appendix B
Approximate Formula for $P_{\delta\theta}$ and $P_{\theta\theta}$ for Arbitrary Redshift

Equation 4.19 in this thesis relates $P_{xy}(z') - P_{\delta\delta}(z')$ at $z = z'$ to the same expression at redshift $z = 0$ using a variable c^2. Note from Eq. 4.16 $g(P_{\delta\delta}(z = 0)) = P_{xy}(z = 0)$ in Eq. 4.19. From Eqs. 4.17 in Chap. 4 and using the result by Scoccimarro et al. 1998 we can write the following solutions for θ and δ in terms of scalings of the initial density field (Bernardeau et al. 2002),

$$\theta(z) = D(z)\theta_1 + D^2(z)\theta_2 + D^3(z)\theta_3 + \cdots \tag{B.1.1}$$

and

$$\delta(z) = D(z)\delta_1 + D^2(z)\delta_2 + D^3(z)\delta_3 + \cdots . \tag{B.1.2}$$

Squaring these expressions and ensemble averaging we can write the velocity divergence power spectrum and the matter power spectrum to third order in perturbation theory as

$$P_{\theta\theta}(z') \sim \langle |D(z')\theta_1 + D^2(z')\theta_2 + D^3(z')\theta_3|^2 \rangle \tag{B.1.3}$$

$$P_{\delta\delta}(z') \sim \langle |D(z')\delta_1 + D^2(z')\delta_2 + D^3(z')\delta_3|^2 \rangle . \tag{B.1.4}$$

Using the fact that $|D\theta_1 + D^2\theta_2 + D^3\theta_3| \leq |D\theta_1| + |D^2\theta_2| + |D^3\theta_3|$ we can approximate this as

$$P_{\theta\theta}(z') \leq \langle (D(z')|\theta_1| + D^2(z')|\theta_2| + D^3(z')|\theta_3|)^2 \rangle \tag{B.1.5}$$

$$P_{\delta\delta}(z') \leq \langle (D(z')|\delta_1| + D^2(z')|\delta_2| + D^3(z')|\delta_3|)^2 \rangle , \tag{B.1.6}$$

and we assume that

E. Jennings, *Simulations of Dark Energy Cosmologies*, Springer Theses, DOI: 10.1007/978-3-642-29339-9, © Springer-Verlag Berlin Heidelberg 2012

$$\langle |D(z')\theta_1 + D^2(z')\theta_2 + D^3(z')\theta_3|^2 \rangle$$
$$- \langle (D(z')|\theta_1| + D^2(z')|\theta_2| + D^3(z')|\theta_3|)^2 \rangle \sim$$
$$\langle |D(z')\delta_1 + D^2(z')\delta_2 + D^3(z')\delta_3|^2 \rangle$$
$$- \langle (D(z')|\delta_1| + D^2(z')|\delta_2| + D^3(z')|\delta_3|)^2 \rangle. \tag{B.17}$$

Taking the difference of the two power spectra we have

$$P_{\theta\theta}(z') - P_{\delta\delta}(z') \sim$$
$$\langle (D(z')|\theta_1| + D^2(z')|\theta_2| + D^3(z')|\theta_3|)^2 \rangle$$
$$- \langle (D(z')|\delta_1| + D^2(z')|\delta_2| + D^3(z')|\delta_3|)^2 \rangle \tag{B.18}$$

and as $x^2 - y^2 = (x - y)(x + y)$ we can rewrite this as

$$P_{\theta\theta}(z') - P_{\delta\delta}(z') \sim$$
$$\langle [D(|\theta_1| - |\delta_1|) + D^2(|\theta_2| - |\delta_2|) + D^3(|\theta_3| - |\delta_3|)]$$
$$\times [D(|\theta_1| + |\delta_1|) + D^2(|\theta_2| + |\delta_2|) + D^3(|\theta_3| + |\delta_3|)] \rangle. \tag{B.19}$$

Multiplying out the rhs of this equation and denoting the modulus of variable $|x|$ as x for simplicity, we have

$$P_{\theta\theta}(z') - P_{\delta\delta}(z') \sim$$
$$\langle \{D^2[\theta_1^2 - \delta_1^2] + D^3[(\theta_1 - \delta_1)(\theta_2 + \delta_2) + (\theta_1 + \delta_1)(\theta_2 - \delta_2)]$$
$$+ D^4[(\theta_1 - \delta_1)(\theta_3 + \delta_3) + (\theta_2^2 - \delta_2^2) + (\theta_1 + \delta_1)(\theta_3 - \delta_3)]$$
$$+ D^5[(\theta_2 - \delta_2)(\theta_3 + \delta_3) + (\theta_2 + \delta_2)(\theta_3 - \delta_3)] + D^6[\theta_3^2 - \delta_3^2]\} \rangle, \tag{B.1.10}$$

and then taking out a factor of $[\theta_1^2 - \delta_1^2]$ on the rhs we have

$$P_{\theta\theta}(z') - P_{\delta\delta}(z') \sim$$
$$\langle [\theta_1^2 - \delta_1^2]\{D^2 + D^3[\frac{\theta_2 + \delta_2}{\theta_1 + \delta_1} + \frac{\theta_2 - \delta_2}{\theta_1 - \delta_1}]$$
$$+ D^4[\frac{\theta_3 + \delta_3}{\theta_1 + \delta_1} + \frac{\theta_2^2 - \delta_2^2}{\theta_1^2 - \delta_1^2} + \frac{\theta_3 - \delta_3}{\theta_1 - \delta_1}]$$
$$+ D^5[2\frac{\theta_3\theta_2 - \delta_3\delta_2}{\theta_1^2 - \delta_1^2}] + D^6[\frac{\theta_3^2 - \delta_3^2}{\theta_1^2 - \delta_1^2}]\} \rangle. \tag{B.1.11}$$

As θ_1 and δ_1 are linear in the initial density contrast, which we assume to be different to the linear density contrast, $\theta_1 \sim \delta_1 \sim \delta_i$ and $\theta_2 \sim \delta_2 \sim \delta_i + \delta_i^2$ is quadratic in the initial density contrast and $\theta_3 \sim \delta_3 \sim \delta_i + \delta_i^2 + \delta_i^3$ is cubic in the initial density field, we assume $\theta_1 + \theta_2 \sim \delta_1 + \delta_2$, $\theta_1 + \theta_3 \sim \delta_1 + \delta_3$ and $\theta_1 - \theta_2 \sim \delta_1 - \delta_2$, $\theta_1 - \theta_3 \sim \delta_1 - \delta_3$ so the fractions in the above equation are unity and

$$P_{\theta\theta}(z') - P_{\delta\delta}(z')$$
$$\sim \langle [\theta_1^2 - \delta_1^2] \rangle \{ D^2 + 2D^3 + 3D^4 + 2D^5 + D^6 \}$$
$$\sim \langle [\theta_1^2 - \delta_1^2] \rangle \{ D(z') + D^2(z') + D^3(z') \}^2 \qquad (B.1.12)$$

Similarly for $P_{\theta\theta}(z) - P_{\delta\delta}(z)$ we have

$$P_{\theta\theta}(z) - P_{\delta\delta}(z)$$
$$\sim \langle [\theta_1^2 - \delta_1^2] \rangle \{ D(z) + D^2(z) + D^3(z) \}^2 \qquad (B.1.13)$$

Taking the ratio of the two previous equations, the redshift independent factor $[\theta_1^2 - \delta_1^2]$ cancels and we obtain the following ansatz

$$\frac{P_{\theta\theta}(z') - P_{\delta\delta}(z')}{P_{\theta\theta}(z) - P_{\delta\delta}(z)} \sim \frac{[D(z') + D(z')^2 + D(z')^3]^2}{[D(z) + D(z)^2 + D(z)^3]^2} \qquad (B.1.14)$$

which is the expression in Eq. 4.19 in this thesis for $z = 0$. A similar approximation works for the cross power spectrum $P_{\delta\theta}$.

References

Komatsu E. et al (2009). ApJS 180:330

Sánchez A.G., Crocce M., Cabré A., Baugh C.M., Gaztañaga E. (2009). MNRAS 400:1643

Percival W.J., et al. (2007). MNRAS 381:1053

Eisenstein D.J., Hu W. (1998). ApJ 496:605

Kowalski M., et al. (2008). ApJ 686:749

Hu W., Sugiyama N. (1996). ApJ 471:542

Bond J.R., Efstathiou G., Tegmark M. (1997). MNRAS 291:L33

Gaztanaga E., Cabre A., Hui L., (2008). arXiv:0807.3551

Eisenstein D.J., et al., (2005). ApJ 633:560

Guy J., Astier P., Nobili S., Regnault N., Pain R. (2005). A&A, 443:781

Li H., Xia J.-Q., Zhao G.-B., Fan Z.-H., Zhang X. (2008). ApJ 683:L1

Corasaniti P.S., Kunz M., Parkinson D., Copeland E.J., Bassett B.A. (2004). Phys. Rev. D 70:083006

Bernardeau F., Colombi S., Gaztanaga E., Scoccimarro R. (2002). Phys. Rept 367:1